无损检测专业英语

主　编　曹　艳　喻星星
副主编　张咏军　朱　颖　唐圣峰
　　　　王志敏　赵迎春　都昌兵

北京理工大学出版社
BEIJING INSTITUTE OF TECHNOLOGY PRESS

内 容 简 介

本书包含7个主要项目，依次是无损检测技术与质量管理、磁粉检测技术、渗透检测技术、目视检测技术、射线检测技术、超声检测技术、涡流检测技术，同时在书中又穿插介绍了新型无损检测技术。针对常用的六种无损检测技术，重点从检测原理、检测设备、方法分类、检测步骤、检测应用、检测工艺等方面进行介绍。每一个部分都编写了对应的习题，考核学习效果，精选原理图片、强调行业标准、强化应用情境与检测任务考核。

本书可供理化测试与质检技术（无损检测方向）专业学生使用，也可作为无损检测及相关专业的工程技术人员及管理人员使用的参考教材或系统培训教材。

版权专有　侵权必究

图书在版编目（CIP）数据

无损检测专业英语 / 曹艳，喻星星主编. --北京：北京理工大学出版社，2022.4
ISBN 978-7-5763-1241-6

Ⅰ.①无… Ⅱ.①曹… ②喻… Ⅲ.①无损检验—英语 Ⅳ.①TG115.28

中国版本图书馆CIP数据核字（2022）第059300号

出版发行 / 北京理工大学出版社有限责任公司
社　　址 / 北京市海淀区中关村南大街5号
邮　　编 / 100081
电　　话 / (010) 68914775（总编室）
　　　　　 (010) 82562903（教材售后服务热线）
　　　　　 (010) 68944723（其他图书服务热线）
网　　址 / http://www.bitpress.com.cn
经　　销 / 全国各地新华书店
印　　刷 / 河北鑫彩博图印刷有限公司
开　　本 / 787毫米×1092毫米　1/16
印　　张 / 20　　　　　　　　　　　　　　责任编辑 / 阎少华
字　　数 / 484千字　　　　　　　　　　　　文案编辑 / 阎少华
版　　次 / 2022年4月第1版　2022年4月第1次印刷　责任校对 / 周瑞红
定　　价 / 89.00元　　　　　　　　　　　　责任印制 / 边心超

图书出现印装质量问题，请拨打售后服务热线，本社负责调换

前　言

本书是一本为了培养具备一定专业英语基础的技术技能型人才，促进无损检测专业教学和培训的系统化而编写的活页式教材。本书采用项目式体例，分别介绍各类无损检测方法的原理、设备、方法、工艺、应用等，学员可根据实际检测项目选取对应的项目进行学习、测验与拓展。

本书重点培养无损检测学员查阅外文专业资料，掌握国外无损检测技术与工艺，与国外同行进行技术交流与合作，全面提升查询、阅读、书写等专业英语的能力，以使其获得必备的外语专业词汇与句型等知识，提升职业素养，与国际接轨。

全书包含7个主要项目，依次是无损检测技术与质量管理、磁粉检测技术、渗透检测技术、目视检测技术、射线检测技术、超声检测技术、涡流检测技术，同时在书中又穿插介绍了新型无损检测技术。针对常用的六种无损检测技术，重点从检测原理、检测设备、方法分类、检测步骤、检测应用、检测工艺等方面进行介绍。每一个部分都编写对应的习题，考核学习效果，精选原理图片、强调行业标准、强化应用情境与检测任务考核。

本书可作为理化测试与质检技术（无损检测方向）专业教材，也可作为无损检测及相关专业的工程技术人员及管理人员使用的参考教材或系统培训教材。

本书具有以下特色：

1. 立足无损检测职业岗位特色，为学生提供实用的外文专业标准与规范。

2. 所有的项目都可采用活页式教材、项目式教学。学习人员也可根据自己的工作需要选择对应的内容学习。

3. 依托丰富的图片资源与网络平台，拓展学习内容，并及时更新资源。

4. 应用英语无损检测技术应用情境，将知识点融入企业招聘信息、仪器使用说明书、检测报告实例等。

5. 每一个项目都对重点、难点进行练习与考核，做到教学效果可评可测。

6. 引入职业素养要求与典型检测实施案例，进行课程思政的渗透。

7. 充分体现对接岗位、融入标准、以学生为中心的高职教育特色。

本书由长沙航空职业技术学院曹艳、喻星星担任主编；由西安航空职业技术学院张咏军，江西南昌航空学校朱颖，特种设备无损检测行业协会专家唐圣峰和长沙航空职业技术学院王志敏、赵迎春、都昌兵担任副主编。

由于编者水平有限，编写时间仓促，书中难免存在不当和疏忽之处，恳请各位同仁指正。

编　者

目 录 Contents

01 Item 1 NDT & Quality Control /1

1.1 Definition, Classification & Criteria for NDT ⋯ 1
1.2 Non-destructive Testing Personnel Certification & Assessment ⋯ 11
1.3 Applications of Non-destructive Testing ⋯ 18

02 Item 2 Magnetic Partical Testing /28

2.1 Theory of Magnetic Partical Testing ⋯ 28
2.2 Classification of Magnetic Particle Testing ⋯ 39
2.3 Instruments and Equipment of Magnetic Particle Testing ⋯ 46
2.4 Magnetic Particle Testing Procedure ⋯ 56

03 Item 3 Liquid Penetrant Testing /76

3.1 Theory of Penetrant Testing ⋯ 76
3.2 Method of PT ⋯ 84
3.3 Instruments and Equipment of Penetrant Testing ⋯ 91
3.4 The Penetrant Progress ⋯ 96

04 Item 4 Visual Testing /116

4.1 Introduction of Visual Testing ⋯ 116

4.2 Theory and Principles ·········· 122
4.3 Welding Terms ·········· 130
4.4 Industrial Endoscope ·········· 137

05 Item 5 Radiographic Testing /149

5.1 Theory of Radiographic Testing ·········· 149
5.2 Physics of Radiography ·········· 156
5.3 Principles of X-radiography ·········· 161
5.4 Equipment & Materials ·········· 168
5.5 Techniques & Calibrations ·········· 174

06 Item 6 Ultrasonic Testing /204

6.1 Introduction of Ultrasonic Testing ·········· 204
6.2 Physics of Ultrasound ·········· 211
6.3 Transducers and Other Equipment ·········· 224
6.4 Measurement Techniques ·········· 239
6.5 Some Applications ·········· 250

07 Item 7 Eddy Current Testing /264

7.1 Introduction of Eddy Current Testing ·········· 264
7.2 Instrumentation ·········· 269
7.3 Probe/Coil Design ·········· 274
7.4 Procedure Issues ·········· 281
7.5 Applications ·········· 287

References ·········· 314

01 NDT & Quality Control

Learning Objectives

1. Knowledge objectives

(1) To grasp the words, related terms and abbreviations about NDT.

(2) To grasp the classification and criteria about NDT.

(3) To know the main applications of NDT.

2. Competence objectives

(1) To be able to read and understand frequently used & complex sentence patterns, capitalized English materials and obtain key information quickly.

(2) To be able to communicate with English speakers about the topic freely.

(3) To be able to fill in the job cards in English.

3. Quality objectives

(1) To be able to self-study with the help of aviation dictionaries, the Internet and other resources.

(2) To train non-destructive testing personnel's professional goal literacy.

1.1 Definition, Classification & Criteria for NDT

〖Point 1〗 Definition

NDT is the English abbreviation for non-destructive testing.

NDT refers to the implementation of a material or workpiece without harming or affecting its future performance or use. By using NDT, we can find the defects in the interior and surface of the material or workpiece, measure the geometric characteristics and dimensions of the workpiece, and measure the internal composition, structure, physical properties and state of the material or workpiece.

What's NDT?

The field of NDT is a very broad, interdisciplinary field that plays a critical role in assuring that structural components and systems perform their function in a reliable and cost effective fashion. NDT technicians and engineers define and implement tests that locate and characterize material conditions and flaws that might otherwise cause planes to crash, reactors to fail, trains to derail, pipelines to burst, and a variety of less visible, but equally troubling events. These tests are performed in a manner that does not affect the future usefulness of the object or material. In other words, NDT allows parts and material to be inspected and measured without damaging them. Because it allows testing without interfering with a product's final use, NDT provides an excellent balance between quality control and cost-effectiveness. Generally speaking, NDT applies to industrial testings. Technology that is used in NDT is similar to those used in the medical industry; yet, typically nonliving objects are the subjects of the testings.

What is NDE?

NDE (Non-destructive evaluation) is a term often used interchangeably with NDT. However, technically, NDE is used to describe measurements that are more quantitative in nature. For example, an NDE method would not only locate a defect, but it would also be used to measure something about that defect such as its size, shape and orientation. NDE may be used to determine material properties, such as fracture toughness, formability and other physical characteristics.

Some NDT/NDE Technologies:

Many people are already familiar with some of the technologies that are used in NDT and NDE from their uses in the medical industry. Most people have also had an X-ray taken and many mothers have had ultrasound used by doctors to give their baby a checkup while still in the womb. X-rays and ultrasound are only a few of the technologies used in the field of NDT/NDE. The number of testing methods seems to grow daily, but a quick summary of the most commonly used methods is provided below.

■ 〖Point 2〗 Common NDT Methods & Their Abbreviations

NDT contains many methods that have been effectively applied (Fig. 1–1). The most commonly used NDT methods are radiographic detection, ultrasonic detection, eddy current detection, magnetic particle detection, penetration detection, visual detection, leakage detection, acoustic emission detection, radiographic fluoroscopy detection, etc. Since various NDT methods have their own scope and limitations, new NDT methods have been continuously developed and applied. Typically, any physical, chemical, or other possible technical means that meet the basic definition of NDT may be developed as a NDT method.

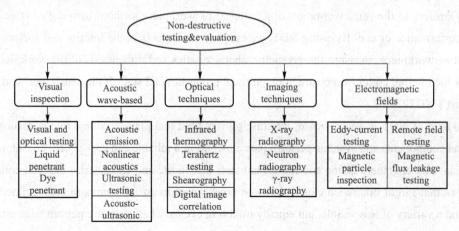

Fig. 1-1 Categories of Non-Destructive Testing & Evaluation Techniques

Ultrasonic testing (UT) (Fig. 1-2) is carried out using sound waves. Sound moves in a straight line in metal. At a transition between two materials sound waves are reflected. This phenomenon is used to detect irregularities. All materials that can be penetrated by ultrasonic waves are suitable for this method: metal, aluminum, copper, stainless steel, etc.

Ultrasonic testing enables faults deep in the material to be detected in different materials over their whole thickness, without having to carry out destructive tests. Ultrasonic testing is also faster than radiography.

Applications: wall thickness measurements, crack detection, weld examination, detection of corrosion/erosion.

 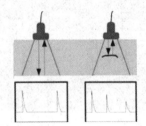

Fig. 1-2 Ultrasonic Testing

Magnetic testing (MT) (Fig. 1-3) makes it possible to identify faults on and directly below the surface. This can only be done in magnetizable grades of steel. Material is tested using a magnetic field and test ink. Cracks, binding faults etc. interfere with the magnetic field.

Applications: surface-examination of welds, high-pressure steam piping, welding of supports, lifting lugs, stay rope attachments, etc.

Fig. 1-3 Magnetic Testing

Penetrant testing (PT) (Fig. 1-4) is a surface-examination that can be carried out on all non-porous materials. A liquid—a "penetrant"—is applied to the material surface. This penetrant then penetrates through into the damage. Another substance—the developer—shows where the penetrant has been absorbed into the damage.

Applications: all non-porous materials, surface testing of welds new build or maintenance, welding of supports, lifting lugs, stay rope attachments, etc.

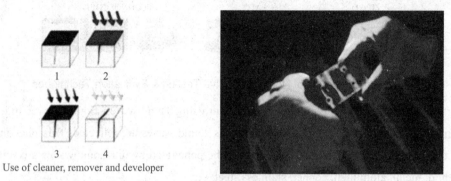

Fig. 1-4　Penetrant Testing

Radiographic Testing (RT) (Fig. 1-5) is the most commonly known non-destructive test method. Radiography can be used to obtain permanent image of surface and sub-surface (embedded) discontinuities.

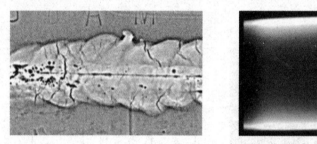

Fig. 1-5　Radiographic Testing

The same discontinuities can be radiographed again after a period of service life and the radiographs can be compared to measure the change in the size and shape of the discontinuity.

There are numerous applications (Fig. 1-6) of radiography in engineering applications. Some of the common uses are detection of surface and subsurface features of interest in welded parts, castings, forgings, wall thickness measurement, corrosion mapping, detection of blockages inside sealed equipment, detection of reinforcing material in concrete slabs, measuring bulk density of materials, measuring porosity in concrete, etc.

Notes

　　Ultrasonic Testing—UT 超声检测

　　Magnetic Particle Testing—MT 磁粉检测

　　Computed Tomographic Testing—CT 计算机层析成像检测

　　Visual Testing—VT 目视检测

　　Radiographic Testing—RT 射线照相检测

　　Penetrant Testing—PT 渗透检测

Acoustic Emission Testing—AT/AE 声发射检测
Eddy Current Testing—ET 涡流检测
Leak Testing—LT 泄漏检测

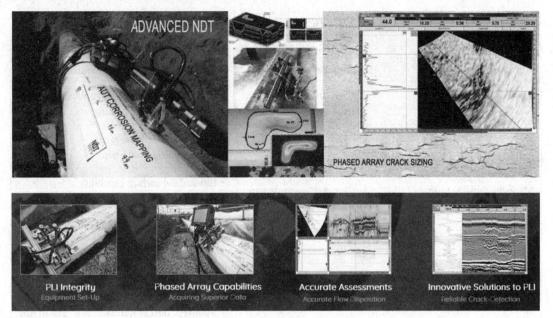

Fig. 1-6 Numerous Applications

For example, A350 is a new generation of wide-body aircraft from Airbus and is the most advanced and efficient wide-body aircraft series in the world today. It is characterized by the large-scale use of advanced materials, 70% of which are made of composite materials (53%), titanium and a new generation of aluminum alloys and other advanced materials. The proportion of the NDT method used by the A350 (Fig. 1-7).

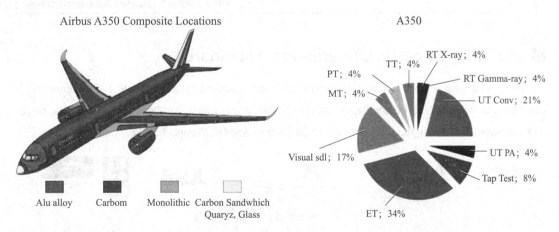

Fig. 1-7 NDT Method Used by A350

The commonly used NDT methods (Table 1-1)are Ultrasonic Scanning, Eddy Current, Dye Penetrant, Magnetic Particle Testing, Microscopy, Acoustic Emission Monitoring, X-ray Radiography, Visual Inspection, Infrared Testing. Other non-destructive detection methods are: Thermal/Infrared

(Tir), AC Field Measurement Technology (Acfmt), Leakage Magnetic Test (Mfl), Long Field Test Detection Method (Rft), Diffraction Wave Time Difference Ultrasonic Detection Technology (Tofd Abbreviation), Wave Guide Detection (Gwt).

Table 1-1 Commonly Used NDT Methods & Limitations

Method	Capabilities	Limitations
Ultrasonic scanning	Subsurface damage	· Material must be good conductor of sound
Eddy current	Surface and near-surface damage	· Hard to analyze in some applications · Just for tubing and coating
Dye penetrant	Surface damage	· No subsurface damage · Not for porous materials
Magnetic particle testing	Surface and layers damage	· Limited subsurface capability · Only for ferromagnetic materials
Microscopy	Small surface damage	· Not suitable for large components · No subsurface damage
Acoustic emission monitoring	Can analyze whole sample	· Hard to analyze · Expensive equipment
X-ray radiography	Subsurface damage	· Smallest defect detectable is 2% of the thickness · Radiation protection · No subsurface damage · Not for porous materials
Visual inspection	Surface damage	· Small damage is difficult to detect · No subsurface damage
Infrared testing	Surface defect	· Not suitable for thick samples · Not suitable for samples that can reflect heat

〖Point 3〗 Domestic & International Standards

NDT, as a widely used testing technology, has clear standard systems (Table 1-2). Generally, it can be divided into domestic standars (Table 1-3) and international standards (Fig. 1-8). The major NDT standard systems in the world are the United States ASME, Europe and ISO standards.

Fig. 1-8 International Standards

Founded in 1880, ASME (American Association of Mechanical Engineers) is an international academic organization with great authority and influence with branches around the world. ASME is mainly engaged in the development of science and technology in mechanical engineering and its related fields, encouraging basic esearch, promoting academic exchanges, developing cooperation with other engineering and associations, carrying out standardization activities, and formulating mechanical norms and standards.

EN standard is based on the common obligations of the participating States, through which the relevant national standard EN a member State will be given legal status or the relevant standard of a State against which it is opposed will be withdrawn, i.e. the national standards of the member states must be consistent with the EN standards.

ISO (International Organization for Standardization) is standard of the International Organization for Standardization, that is, the standard formulated by the International Organization for Standardization. ISO is an International Organization for Standardization (ISO) with members from national standardization groups from more than 100 countries around the world.

Table 1-2　Assessment Criteria and Level of Acceptance for Training

Standards	Acceptance Level
JB 4730-94（部标）	Ⅰ级，Ⅱ级，Ⅲ级
GB/T 2970-91（国标）	B4级，B3级，B2级
BS 5996: 1993（英国）	Ⅰ级，Ⅱ级，Ⅲ级
SEL 072-79（德国）	Ⅱ级，Ⅲ级，Ⅳ级
JISG 0801-1993（日本）	合格
NFA 04-305 (1983)（法国）	B级，A级
ASTM 435/A435M-90（美国）	合格
ASTMA 578/A578M-96（美国）	CBA
EN 10160-1999（欧洲）	S2E2, S1E2, S0E2

Table 1-3　Domestic NDT Standards

Standard Type	Standard	Content
National Standards	GB/T 5616-2014	Non-destructive testing application guidelines
	GB/T 5677-2007	Radiographic examination of cast steel
	GB/T 9443-2007	Permeation detection of cast steel
	GB/T 9444-2007	Magnetic particle testing of cast steel
	GB/T 9445-2015	Non-destructive testing personnel qualification and certification
National Military Standards	GJB 593.1-1988	Non-destructive testing quality control specification for ultrasonic longitudinal and shear wave testing
	GJB 593.2-1988	NDT quality control specification X-radiographic testing
	GJB 593.3-1988	Non-destructive testing quality control specification for magnetic powder testing
	GJB 593.4-1988	Non-destructive testing quality control specification penetration testing

continued

Standard Type	Standard	Content
Ship Industry Standards	CB/T 3177-1994	Rules for radiography and ultrasonic testing of steel welds in ships
	CB/T 3290-2013	Permeation detection of copper alloy propeller in civil ships
	CB/T 3558-2011	Process and quality classification of marine steel seam X-ray testing
	CB/T 3559-2011	Ultrasonic testing technology and quality classification of ship steel welds
	CB/T 3802-1997	Testing requirements for hull weld surface quality
Electricity Industry Standards	DL/T 664-2016	Application specification for infrared diagnosis of live equipment
	DL/T 675-2014	Qualification rules for non-destructive testing personnel in power industry
	DL/T 694-2012	Technical guidelines for ultrasonic testing of high temperature fastening bolt
	DL/T 714-2011	Technical guidelines for ultrasonic testing of steam turbine blade
Aviation Industry Standards	HB 5169-1981	Ultrasonic testing method for platinum-iridium-25 alloy plate
	HB 5265-1983	Ultrasonic testing instructions for TC11 titanium alloy compressor plates and cake (ring) billets for aeroengine
	HB 5266-1983	Acceptance criteria for ultrasonic testing of TC11 titanium alloy compressor plates and cake (ring) billets for aeroengine
	HB 5356-1986	Vortex test method for conductivity of aluminum alloy
Energy Industry Standards	NB/T 20236-2013	Non-destructive testing of weld seam in steel lining of PWR nuclear power plant
	NB/T 20243-2013	Ultrasonic examination of damaged fuel assemblies in PWR
	NB/Z 20254-2013	Nuclear power plant reactor coolant system leak detection guidelines

Put into Practice

1. Read the following recruitment information and answer what responsibilities and requirements the job have?

Responsibilities:

(1) Responsible for on-site NDT project management, including organize and manage NDT inspector team, monitoring NDT performing to meet client and company requirement.

(2) Team leading.

(3) HSE control.

(4) Project document control.

(5) Communication with client and problem solving.

(6) Team training.

Requirements:

(1) At least 5 years of experience in on-site NDT project. Level II NDT certificates on RT/UT/MT/PT, Level III is preferred.

(2) Experience in team management, good leadership.

(3) Experience in Shipyard or offshore engineering is preferred.

(4) Good communication skills.

(5) Reliable and integrity.

(6) Process PAUT certificate, and PAUT field testing experience.

2. Read the following training information from Table 1-4 and answer the questions below. What training does the school carry out? What are the standards of these training? What is the training plan for 2021?

Birring NDT Training—MT, PT, UT, VT, ET, PAUT, RT.

Birring NDE Center, Inc. provides NDT Level II training in Houston. NDT training as per SNT-TC-1A or NAS-410. Methods includes magnetic particle testing, liquid penetrant testing, ultrasonic testing, phased array testing, eddy current testing, radiographic testing film interpretation and radiation safety.The school is approved by the Texas Workforce Commission (TWC) . Birring NDE Center is approved to train Veterans. Birring NDE Center, Inc. is an excellent school for Veterans for NDT Training. VA Facility Code is 25149143. Students are eligible for financial aid.

Table 1-4 2021 SCHEDULE

DATE	MON	TUE	WED	THU	FRI	SAT	SUN
Jan 4-8		MT I & II		PT I & II			
Jan 4-8		Radiographic Testing I					
Jan 11-15		Ultrasonic Testing I					
Jan 11-15		Eddy Current Testing I					
Jan 18-22		Ultrasonic Testing II					
Jan 18-22		Eddy Current Testing II					
Jan 25-27		VT I & II					
Feb 1-5		MT I & II		PT I & II			
Feb 8-12		Ultrasonic Testing I					
Feb 8-12		Eddy Current Testing I					
Feb 8-12		Radiographic Testing I					

continued

2021 SCHEDULE							
DATE	MON	TUE	WED	THU	FRI	SAT	SUN
Feb 15-19		Ultrasonic Testing Ⅱ					
Feb 15-19		Eddy Current Testing Ⅱ					
Feb 22-24		VTⅠ & Ⅱ					

3. Read the following passage and answer what are the testing items?

Testing Service

Oil storage detection, on-line acoustic emission testing of gas storage tank, PEC detection, scanning test of tank floor (internal and external corrosion and leak detection) eddy current testing and conventional RT/UT/MT/PT testing. Operation scope including radiographic testing of boilers, spherical tank and other pressure vessels, pressure piping and special equipment, non-destructive testing of large axial parts, bolts, screws, steel structure material weld, 500 sets of safety assessment tank (200-100 000 cubic meters), 500 000 of X-rays, shooting; 200 000 m detection of ultrasonic, magnetic powder, permeation, with tank detection and testing as the development direction, the boiler, pressure vessel professional subcontracting and detection of gas pipeline, petrochemical plant, chemical plant.

Detection scope

◆ Radiographic testing of boiler, pressure vessel, pressure piping and special equipment, testing technology of ultrasonic testing, magnetic particle testing, penetrant testing and acoustic emission

◆ Metal test of large power plant construction project, metal supervision and inspection of power plant repair

◆ Conventional eddy current testing of ferromagnetic materials and ferromagnetic materials heat exchanger tube and the far field eddy current testing technology

◆ Physical and chemical, metallurgical, spectrum analysis technology of metal materials

◆ Non-destructive testing technology of steel structure material and weld

◆ Ultrasonic and magnetic particle testing technology of large axial parts, bolt and screw parts

◆ Acoustic emission testing technology of the spherical tank, oil tank, urea synthesis tower, long tube trailer

◆ X-ray real-time imaging detection, phased array and TOFO detection technology

◆ Ultrasound B and C scan imaging technology

◆ High temperature ultrasonic testing technology of power plant

◆ Personnel training and further education of non-destructive testing, ASNT Ⅰ, Ⅱ level personnel training assessment

◆ Non-destructive testing data and standard queries

◆ Technical consultation of related non-destructive testing problems

◆ Consulting of welding technique problems such as welding procedure qualification
◆ Mechanical property test and chemical component analysis
◆ Pipe, tube plate butt weld test (small diameter tube)

1.2 Non-destructive Testing Personnel Certification & Assessment

■ 〖Point 1〗 Non-destructive Testing Qualification

1. Non-destructive testing personnel qualification (Fig. 1-9) rating is divided into three levels:

Class I holders shall, under the supervision or guidance of level II or III personnel, carry out NDT according to the established NDT operating instructions, and, under the capabilities specified in their certificates, with the authorization of the employer, Class I personnel may perform the following tasks in accordance with the NDT operating instructions:

(1) Adjust NDT equipment;

(2) Perform tests;

(3) Record and classify test results;

(4) Report the test results, but can not evaluate the test results.

2. Class II holders shall have the ability to perform NDT in accordance with established process procedures:

(1) Optional detection method;

(2) Prepare NDT operating instructions;

(3) Adjustment and calibration equipment;

(4) Interpret and evaluate test results in accordance with applicable specifications, standards, technical conditions or process procedures;

(5) Prepare NDT test reports;

(6) Implement and supervise and guide level II or I testing.

3. Class III holders shall be of the highest level of qualification and shall have the ability to perform and direct NDT operations, within the specified in their certificates:

(1) Develop and validate NDT operating instructions and process procedures to review their editorial and technical integrity;

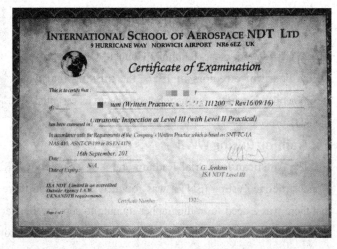

Fig. 1-9 Non-destructive Testing Personnel Qualification

(2) Interpretation of specifications, standards, technical conditions and procedures;

(3) Implement and supervise the testing of personnel at all levels;

(4) Determine the specific testing methods, process procedures and NDT operating instructions used.

Training and Experience Requirements to be a NDT Level II as per SNT-TC-1A 2016 (Fig. 1-10). Example: To be a MT Level II, an individual must have 20 hours of class room training, 280 hours of MT experience and 530 hours of total NDT experience.

Type	PT	MT	VT	ET or UT
Training	12 hours	20 hours	24 hours	80 hours
Experience-Method	210 hours	280 hours	210 hours	840 hours
Experience-Total NDT	400 hours	530 hours	400 hours	1 600 hours
NOTE: Certification of all levels of NDT is the responsibility of the employer (9.1 of SNT-TC-1A)				

Fig. 1-10 Training Hours

■ 〖Point 2〗 International NDT Training

The Evolution of NDT Training

The need for training has always been essential to the growth of NDT. The technology of non-destructive testing has experienced phenomenal growth over the past two decades and it is expected that this growth will continue for the foreseeable future. The need for training has always been essential to this growth but there was a time when this was not so. In the early days of NDT, training was mostly unorganized and consisted of on-the-job efforts. An experienced practitioner would pass on their skills to a "helper" by the "show and tell" method. There was little to help this effort in the way of training materials. In 1964, a draft document was developed by a Task Group, then approved for publication by the Technical Council for the membership of SNT (prior to the name change to ASNT) to review and comment. That document was referred to as "SNT-TC-1T" (Fig. 1-11); the "T" was an abbreviation for "Tentative". It would ultimately be changed to SNT-TC-1A. The "TC" referred to the Technical Council. Once approved by the board of directors in 1966, the training that was recommended began to seriously be considered as essential to the NDT process.

Fig. 1-11 SNT-TC-1T

Training During WWII

The American Industrial Radium and X-ray Society (AIRXS), the forerunner of the Society for Nondestructive Testing (SNT), was founded in 1941 and there was almost immediately a realization that there was a need for radiographer training to support the war effort. As a result, several colleges and universities established courses in industrial radiography including the Illinois Institute of Technology, University of Minnesota, Century College (Chicago), Marquette University, Massachusetts State

Department of Education, Michigan College of Mining and Technology, Canisius College (Buffalo, NY), University of Illinois, Columbia University School of Mines, and Hillyer College (Hartford, CT). But most beginners had to learn how to be a radiographer from those who already had experience.

The U.S. government also provided some limited training. For example, on March 11, 1943, the War Production Board held a clinic in Chicago dedicated to radiography and promoted other methods including penetrant and magnetic particle testing. It was not much, but it was a beginning.

An early report indicated that Magnaflux Corp. conducted fifty-five training sessions between 1941 and 1945, with some 5 000 in attendance. The War Production Board's clinic was involved with that effort.

During World War II, the Society began to experience significant growth, and several local SNT sections offered informal courses with emphasis on practical applications.

Early Training Efforts

Since NDT training was virtually non-existent, many employers depended on experienced employees to provide training on a limited basis. Some manufacturers of NDT equipment would include minimal training materials with the purchase of their products. Motivated employees could also learn through self-study efforts using the training materials available at that time. Those early courses offered by the equipment manufacturers presented the basic principles of the method, but emphasis was placed on the operation and application of the equipment.

ASNT Sections

Short courses were offered by a number of local sections starting in the 1950s. In an early issue of materials evaluation, it was reported that there were over 100 attendees at the Westinghouse Electric building in Philadelphia. The theory, applications, and test results of NDT with emphasis on radiography and ultrasonics were the topics. (Also offered at some of the conferences.)

The Western New York, Chicago, and Boston sections were offering courses in 1955 with others planned for spring 1956.

The Northern California section in cooperation with the American Society for Metals, advertised a ten-week course on radiography and the New York section reported that there were over 60 students attending a class on "industrial radiography" at two different locations.

There were also individual efforts to provide NDT training courses. In 1966, Paul Dick, who was employed by General Electric in King of Prussia, PA, offered evening courses at the Temple University Technical Institute in conjunction with the Philadelphia Section of ASNT.

Within the international scope, NDT standards have certain generality, and the more common NDT qualification standards (such as the European Union's PCN certification based on EN473 standards is universal) are managed by large NDT institutions. And pass in the European Union and the United States and other places.

NDT training also known as NDT training certification, which is a non-destructive testing practitioners for professional training and certification courses, including theoretical training, practical exercises, comprehensive assessment of the whole process. Only qualified NDT societies or authorized institutions can conduct professional NDT training.

There is a huge market in the NDT testing industry, but there are differences in training and certification mechanisms, culture and business practices between Chinese NDT and international NDT. Therefore, both domestic and international NDT training is the responsibility of completely different societies and has different characteristics. The follwing is a NDT personnel qualification record form (Fig. 1-12).

	无损检测Ⅰ&Ⅱ级人员资格 评定记录表 QUALIFICATION RECORD FOR NDE PERSONNEL (LEVEL Ⅰ & Ⅱ)		记录编号 Record No.: _____
代号 ID		姓名 Name	
确定的 NDE 方法 Certified NDE Method		确定的 NDE 方法 Certified NDE Level	
教育 Education	☐大学 University (4 years) ☐大专 College (2 years) ☐技术学校 Technical School (2 years) ☐高中 High School (3 years)		
学科 Type of Degree	☐工科 Engineering ☐理科 Science ☐其他 Other	毕业时间 Date of Graduation	
经历 Experience	见无损检测人员专业工龄证明书 See Experience Certification in NDE of HTAC NDE Personnel		
技能考试 Examination for Technical Qualification	考试项目 Examination Subjects	实际得分 Actual Grade	
	通用考试 General Examination		
	专业考试 Specific Examination		
	实践考试 Practical Examination		
	综合得分 Composite Grade		
视力检查 Vision Examination	☐合格　　Acceptable ☐不合格　Reject	见视力检查记录表 See Vision Examinations Record	
编制 日期 Prepared by Date		审核 日期 Reviewed by Date	
评定结果 Qualification Results: 　　该申请人圆满地完成了培训。他的教育、资历、视力检查及考试均符合 SNT-TC-1A 和_____的要求，鉴定合格有效期从_____至_____。 　　This individual has satisfactorily completed training.His education, qualification, vision examination and grade meet the requirement of SNT-TC-1A and_____. This certificate is valid from _____to_____. 评定人 Qualification Personnel: _____　　日期 Date: _____			

Fig. 1-12　Non-destructive Test (Level I, II) Personnel Qualification Record Form

ASNT-NDT Level-II in RT, UT, MT, PT & VT Course (Fig. 1-13) Details

The non-destructive testing (NDT) are keys for reassuring and controlling the standard of parts and structures throughout producing and in-service inspections. Destructive and non-destructive tests area unit vital a part of quality assurance plans to satisfy client needs. Destructive tests (DT) has narrow applications because the product being tested is broken or destroyed and thence makes tested material rarely helpful.

The use of noninvasive techniques to see the integrity of a materials, components or structure or quantitatively determine some feature of an object, i.e. examine or determine while not doing harm. In today's world wherever new materials are being developed, older materials and bonding ways are being subjected to higher pressures and masses, NDT ensures that materials will still operate to their highest capability with the reassurance that they're going to not fail among planned cut-off dates.

Magnetic Particle Testing course content:
√Introduction
√History of Magnetic Particle Testing
√Magnetism and its properties
√Advantages and limitations
√Principle of magnetic particle testing
√Types of magnetization equipments
√Magnetic particle testing techniques
√Practicals
√Interpretation of indications as per
√Report writing

Ultrasonic Testing course content:
√Introduction
√Principle of Ultrasonic testing
√Advantages and limitations
√Types of waves
√Ultrasonic Testing Techniques
√Thickness measurement
√DAC plotting
√Weld and plate scanning
√Practical in Lab
√Interpretation of indications
√Report writing

Penetrant Testing course content:
√Introduction
√History of Penetrant Testing
√Advantages and limitations
√Principle of penetrant testing
√Penetrant testing process
√Types of penetrants
√Types of developers
√Testing techniques
√Practicals
√Interpretation of indications as per code
√Report writing

Radiography Testing course content:
√Introduction
√History of radiations (x-ray and gamma ray)
√Principle of radiography testing
√Source strength and radiation intensity
√HVT & TVT calculations
√Shadow formation
√X-ray and Gamma ray equipment
√Types of films and screens
√Image quality indicator
√Radiographic Testing Technique
√Film processing
√RTFI
√Report writing
√Practicals with Industrial visit

Fig. 1-13 Assessment Criteria

Future of NDT Training

Online NDT courses became available in the early 2010s and have become extremely popular for many reasons. At first, many potential users were reluctant since it was not generally embraced, but once the benefits and cost savings were realized, it became widely accepted. One of the concerns had to do with the practical (hands-on) lab exercises which is a vital part of NDT training. Along came "blended" training that could be accomplished at the student's company under the guidance of the Level Ⅲ. Also, an instructor could be assigned to oversee the practical training.

There are also efforts being considered to provide remote online practical training. There will be more on this subject as it becomes available.

〖Point 3〗 NDT Code of Ethics

1. Purpose

To define the code of ethics for NDT personnel.

2. Applicability

To all personnel involved in NDT.

3. Requirements

Personnel shall:

(1) Have a responsibility to safeguard the life, health, property, and welfare of the public and to maintain the integrity and high standards of skill and practice in the profession of NDT.

(2) Be aware of and uphold the provisions of all codes, standards, and regulations under which they are working.

(3) Undertake to perform NDT duties only when certified in the specific methods and techniques involved.

(4) Be objective in any NDT Testing, report or statement, avoiding any omission that would or could lead to misrepresentation. Make decisions based upon factual evidence.

(5) Bring concerns to the attention of management/supervision.

Put into Practice

1. Read the training below and indicate what training standards and programs are available.

The South West School of NDT specialises in providing training and examination services (Fig. 1–14) in the fields of NDT and associated technologies.

Fig.1-14 The South West School of Non-destructive Testing

These services are in satisfaction of all major international and European standards:

•ISO 9712

•EN 4179

•NAS 410

- SNT-TC-1A
- CP189
- ATA105

The school also holds independent accreditation via the British Institute of NDT.

Although courses and examinations at all levels (levels Ⅰ, Ⅱ and Ⅲ) are routinely conducted at the training centre based in Cardiff UK, instructors and examiners travel to various parts of the world to provide such services at client locations.

In addition to the main methods of:
- Ultrasonic Testing
- Radiographic Testing
- Eddy Current Testing
- Magnetic Testing
- Penetrant Testing

The school also specialises in structuring courses and examinations in many of the emerging techniques especially those related to composite material inspection.

2. Translate the training materials (Fig. 1–15) below and point out what are the training contents and training levels?

A team of experienced multi-method NDT Level Ⅲ instructors and examiners not only support the training and examination activity but also form the basis of a much used NDT consultancy service. Ranging from an all-encompassing annual consultancy contract to the one-off approval of testing techniques and procedures, members of the team provide advice and technical direction in all facets of NDT. Most serve on national and international committees to assist in the formulation future standards and policy.

Fig.1–15 Training Materials

3. Translate the training materials below and answer some of the questions that should be identified in the NDT career planning?

NDT Industry Career Guide

Welcome to the NDT Training and Testing Center NDT Industry Career Guide. We'll get you started in your NDT Career... The RIGHT WAY!

We know there are lots of questions when you first begin researching the NDT Industry as a viable career option. So, we created this NDT Career Guide to help you get pointed in the right direction.

We have over 40 years of experience in the NDT Industry and the NDT Training Business. Our experienced, well-trained, certified instructors and staff are here to help you learn about this exciting industry.

As you move through the Career guide, we'll answer these questions, and more:
• What is the NDT Industry?
• How do you get into the NDT Industry?
• What type of NDT Training will you need?
• What are the options for obtaining that training?
• How to find a job in the industry?
• What type of jobs are available?
• How to become a certified NDT Technician?

Take the first step toward a successful career in NDT. Choose which way you would like to learn about your NDT Career Opportunities. You'll Be Glad You Did!

Let's Get Started!

1.3 Applications of Non-destructive Testing

〖Point 1〗 History of NDT

Nondestructive testing (NDT), also called non-destructive examination (NDE) and non-destructive inspection (NDI), is testing that does not destroy the test object. NDE is vital for constructing and maintaining all types of components and structures. To detect different defects such as cracking and corrosion, there are different methods of testing available, such as X-ray (where cracks show up on the film) and ultrasound (where cracks show up as an echo blip on the screen). This article is aimed mainly at industrial NDT (Fig. 1-16), but many of the methods described here can be used to test the human body. In fact, methods from the medical field have often been adapted for industrial use, as was the case with phased array ultrasonics and computed radiography.

While destructive testing usually provides a more reliable assessment of the state of the test object, destruction of the test object usually makes this type of test costlier to the test object's owner than non-destructive testing. Destructive testing is also inappropriate in many circumstances, such as forensic investigation. That there is a tradeoff between the cost of the test and its reliability favors a strategy in which most test objects are inspected non-destructively; destructive testing is performed on a sampling of test objects that is drawn randomly for the purpose of characterizing the testing reliability of the non-destructive test.

It is very difficult to weld or mold a solid object that has the risk of breaking in service, so testing at manufacture and during use is often essential. During the process of casting a metal object, for example, the metal may shrink as it cools, and crack or introduce voids inside the structure. Even the best welders (and welding machines) do not make 100% perfect welds. Some typical weld defects that need to be found and repaired are lack of fusion of the weld to the metal and porous bubbles inside the weld, both of which could cause a structure to break or a pipeline to rupture.

Fig. 1-16   Application in Industrial

During their service lives, many industrial components need regular non-destructive tests to detect damage that may be difficult or expensive to find by everyday methods.

For example:

Aircraft skins need regular checking to detect cracks;

Underground pipelines are subject to corrosion and stress corrosion cracking;

Pipes in industrial plants may be subject to erosion and corrosion from the products they carry;

Concrete structures may be weakened if the inner reinforcing steel is corroded;

Pressure vessels may develop cracks in welds;

The wire ropes in suspension bridges are subject to weather, vibration, and high loads, so testing for broken wires and other damage is important.

Over the past centuries, swordsmiths, blacksmiths, and bell-makers would listen to the ring of the objects they were creating to get an indication of the soundness of the material. The wheel-tapper would test the wheels of locomotives for the presence of cracks, often caused by fatigue—a function that is now carried out by instrumentation and referred to as the acoustic impact technique.

Notable Events in Early Industrial NDT

1854, Hartford, Connecticut: a boiler at the Fales and Gay Gray Car works explodes, killing 21 people and seriously injuring 50. Within a decade, the State of Connecticut passes a law requiring annual Testing (in this case visual) of boilers.

1895, Wilhelm Conrad Rontgen discovers what are now known as X-rays. In his first paper, he discusses the possibility of flaw detection.

1880–1920, The "Oil and Whiting" method of crack detection is used in the railroad industry to find cracks in heavy steel parts. (A part is soaked in thinned oil, then painted with a white coating that dries to a powder. Oil seeping out from cracks turns the white powder brown, allowing the cracks to be detected.) This was the precursor to modern liquid penetrant tests.

1920, Dr. H. H. Lester begins development of industrial radiography for metals. 1924, Lester uses radiography to examine castings to be installed in a Boston Edison Company steam pressure power plant.

1926, The first electromagnetic eddy current instrument is available to measure material thicknesses.

1927–1928, Magnetic induction system to detect flaws in railroad track developed by Dr. Elmer Sperry and H.C. Drake.

1929, Magnetic particle methods and equipment pioneered (A.V. DeForest and F.B. Doane).

1930s, Robert F. Mehl demonstrates radiographic imaging using gamma radiation from Radium, which can examine thicker components than the low-energy X-ray machines available at the time.

1935–1940, Liquid penetrant tests developed (Betz, Doane and DeForest).

1935–1940, Eddy current instruments developed (H.C. Knerr, C. Farrow, Theo Zuschlag and Fr. F. Foerster).

1940–1944, Ultrasonic test method developed in USA by Dr. Floyd Firestone.

1950, J. Kaiser introduces acoustic emission as an NDT method.

(Source: Hellier, 2001) Note the number of advancements made during the World War II era, a time when industrial quality control was growing in importance.

■ 〖Point 2〗 The Application Fields of Non-destructive Testing

Without to damage the being detected objects, Non-destructive Testing Techniques can detect surface and internal defects of a variety of engineering materials, components, structural components.

Compared with destructive testing, non-destructive testing has the following characteristics. The first, non-destructive, because it does not damage the performance of the detected object; The second, it is comprehensive, because the detection is non-destructive, so the detected object can be 100% comprehensive detection when necessary, which is not destructive detection; The third, it is full-process, destructive testing is generally only suitable for testing raw materials, such as tensile, compression, bending and so on, which are generally used in mechanical engineering. For finished products and supplies, destructive testing can not be carried out unless they are not prepared to continue in service, and non-destructive testing does not damage the performance of the tested object. Therefore, it can not only the manufacturing raw materials, intermediate process links, until the final finished product for the whole process of testing, but also the equipment in service.

Today, when we talk about non-destructive testing technology, we have extended from the original non-destructive testing limited to machinery, materials, accessories in industry to a wider range of fields such as trade, agriculture, medicine and so on. It is difficult to imagine that non-destructive testing technology has established a very close relationship with food safety, drug testing, fruit testing, planting and cultivation.

NDT technology is widely used in industry (Fig. 1–17), such as aerospace, nuclear industry, weapons manufacturing, machinery industry, ship building, petrochemical, railway and high-speed trains, automobiles, boilers and pressure vessels, special equipment and customs testing.

Non-destructive testing technology not only plays a supervisory role in ensuring food quality and safety in the field of food processing, such as material purchase, quality change of processing process, quality change of circulation link, but also plays an active role in saving energy and raw material resources, reducing production cost, improving finished product rate and labor productivity.

Fig. 1-17 NDT Technology in Industrial Field

Non-destructive testing technology is also playing an irreplaceable role in combating climate change, developing low-carbon economy, circular economy and green remanufacturing industry (Fig. 1-18). Non-destructive testing is an indispensable and effective tool for industrial development, which to some extent reflects the level of industrial development of a country, and its importance has been recognized.

■〖Point 3〗 Purpose of NDT

1. Quality control

The performance and quality level of each product are usually clearly stipulated in its technical documents,

Fig.1-18 The Role of NDT

such as technical conditions, specifications, acceptance standards, etc.which, are characterized by certain technical quality indicators. One of the main purposes of non-destructive testing is to provide real-time quality control of raw materials and components for discontinuous processing (e.g.multi-process production) or continuous processing (e.g.automated production lines), such as control of metallurgical quality of materials, processing process quality, organization status, thickness of coating and size, orientation and distribution of defects, etc.

In the process of quality control, feedback the obtained quality information to the design and process department can, in turn, promote it to further improve the design and manufacturing process of the product, and the quality of the product must be consolidated and improved accordingly, thus reducing the cost and improving the production efficiency. Of course, the use of non-destructive testing technology can also control the quality level of raw materials or products within the scope of design requirements according to acceptance standards, without the need to improve the quality requirements without limit, even without affecting the design performance, use some defective materials to improve the utilization of social resources and improve economic efficiency.

2. In-service testing

The use of non-destructive testing technology to monitor the device or component during operation or to carry out regular testing during the maintenance period, can timely find the hidden trouble that affects the device or component to continue safe operation and prevent the occurrence of accidents. This is of great significance for the prevention of major equipment, such as nuclear reactors, bridge buildings, railway vehicles, pressure vessels, pipelines, aircraft, rockets and so on.

The purpose of in-service detection is not only to detect and identify and eliminate the hidden dangers of the safe operation of the hazardous device in time, but also to evaluate the continued use of the device or component and its safe operation life on the basis of the early defects found and their degree of development (such as the initiation and development of fatigue cracks), on the basis of determining its orientation, size, shape, orientation and properties. Although non-destructive evaluation has just started in China, it has become an important development direction of non-destructive testing technology.

3. Quality appraisal

Final Testing shall be carried out for finished products (including materials, components) before they are assembled or put into use, that is, quality identification. The purpose of this paper is to determine whether the object under testing can achieve the design performance, whether it can be used safely, that is, to judge whether it is qualified or not, which is not only the acceptance of the previous processing procedure, but also to avoid the hidden trouble to the future use. Non-destructive testing techniques are used to detect the conformity of materials or components in each (or one or several) process of casting, forging, welding, heat treatment and cutting to avoid the continued futile processing of unqualified products. This work is generally called quality testing and in essence belongs to the category of quality appraisal. The quality acceptance appraisal before product use is very necessary, especially those products that will be used under complex and harsh conditions such as high temperature, high pressure, high stress, high cycle load, etc. In this respect, non-destructive testing technology shows the incomparable superiority of 100% testing.

To sum up, non-destructive testing technology has shown an extremely important role in production design, manufacturing process, quality appraisal, economic benefit and work efficiency improvement. Therefore, non-destructive testing technology has been increasingly recognized and accepted by visionary business leaders and engineers. The basic theory of non-destructive testing, the detection method and the analysis of the test results, especially the analysis of some typical application examples, have become the necessary knowledge of engineering technicians.

〖Point 4〗 Technological Development

After entering the 21st century, with the development of science and technology, especially computer technology, digitization and image recognition technology, artificial neural network technology and mechatronics technology, non-destructive testing technology has made rapid progress.

Ray imaging and defect automatic recognition techniques, ray computer aided imaging (CR), ray real-time imaging (DR) and ray tomography (CT) have been widely used in ray detection. The rapid X-ray real-time imaging system for detecting containers, various industrial CT devices with X-ray, γ-ray and linear accelerator as ray sources have been widely used in various industrial fields. Microfocus X-ray CT can detect micro defects at micron level.

In ultrasonic detection, various digital ultrasonic flaw detectors are widely used. TOFD ultrasonic detection system, ultrasonic imaging detection system, magnetostrictive ultrasonic guided wave detection system, phased array ultrasonic detection system have been widely used. A large number of research results have been obtained in the research of detection methods and application techniques, mainly aimed at automatic ultrasonic detection technology, ultrasonic imaging detection technology, artificial intelligence and robot detection technology, TOFD ultrasonic detection technology, ultrasonic guided wave detection technology, non-contact ultrasonic technology, phased array ultrasonic detection technology, laser ultrasonic detection technology and so on. A multi-channel ultrasonic flaw detector used in the tube bar and welded pipe automatic detection line has a number of channels up to 500 and a sampling rate up to 240 MHz. Ultrasonic guided wave detection system and magnetostrictive guided wave detection method have been used for long distance detection of corrosion defects in industrial pipelines and buried pipelines with insulation layer.

In the aspect of electromagnetic detection, the conventional eddy current detection instruments are all digitized and the array probe and multi-channel instrument are developed to realize the application of advanced electronic and information technology such as data conversion and analysis. Far-field eddy current, multi-frequency eddy current, pulse eddy current and magneto-optical, eddy current imaging detection technology have been developed and applied. Pulse eddy current detection technology is used for corrosion detection of steel pressure vessels and pipes with insulation layer, which can penetrate 150 mm thick insulation layer.

Magnetic leakage detection technology has been widely used in large atmospheric storage tank bottom plate corrosion detection, pipeline manufacturing process on-line detection, wire rope detection, oil drill pipe detection and no insulation layer industrial pipeline corrosion detection. Magnetic memory detection has been widely used in power plant boilers, pressure vessels, pressure pipes, steam turbines, wind turbines (Fig. 1-19) and bridges. Barkhausen noise technology is more widely used in residual

Fig. 1-19 NDT on Wind Rotor Blades

stress detection.

In acoustic emission detection, various advanced multi-channel acoustic emission instruments are emerging. In the aspect of acoustic emission signal analysis and processing, including conventional parameter analysis, time difference location, correlation graph analysis, spectrum analysis, wavelet analysis, pattern recognition, artificial neural network pattern recognition, fuzzy analysis and gray correlation analysis, etc. In China, there are more than 50 testing institutions engaged in pressure vessel acoustic emission detection.

In microwave detection and infrared detection, has also been greatly developed. Microwave detection is widely used in the detection of humidity, temperature, density, curing degree and so on. It also plays an important role in the detection of adhesive structure, composite material, rocket propellant and so on.

Infrared detection has been widely used in power industry, petrochemical industry, housing construction and other fields. Applied research has also been carried out in the fields of metal mechanics specimen, fracture mechanics and stress analysis, printed circuit board fault analysis and ceramic industry. Infrared thermal imaging of pressure vessel has been formally incorporated into the safety supervision system of special equipment in China.

In-service testing (Fig. 1–20) is an important method of safety monitoring of equipment and structure. It has been widely carried out in the fields of special equipment such as pressure vessel, oil and gas pipeline (Fig. 1–21), aviation system, railway system, civil engineering and steel structure, nuclear power plant and so on, and has made remarkable achievements.

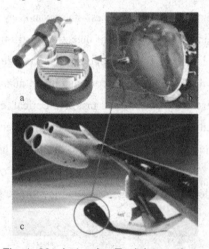

Fig. 1–20　In-service Tank Inspections　　　　Fig. 1–21　Dynamic Excitation on a Spacecraft

The theory and regulations of reliability evaluation of in-service structure have been unanimously recognized internationally. Non-destructive testing technology plays an important role in the reliability evaluation of equipment and structure.

China has 170 000 non-destructive testing personnel and more than 2 000 non-destructive testing institutions (excluding internal), which play an important role in national economic construction and personal and equipment safety monitoring.

Put into Practice

1. Translate the following non-destructive testing service items into Chinese.

NDT Services

CD International Technology, Inc. (CDINT) is led by specialists who have over 30 years of experience in NDT and who hold ASNT level Ⅲ certificates. Our service is available 24 hour per day and 365 days per year. We serve the world market, provide timely and quality inspection services in both traditional and advanced NDT capabilities.

The Fields/Areas We Cover

• Oil and Gas Pipeline

• Refinery

• Petrochemical Plant

• Power Plant (including nuclear power plant)

• Construction

• Aerospace and Aircraft

• Fabrication

• Railroad

2. Read the essay below to answer what services the company provides.

Our NDT Services, LLC is a Non-destructive Testing Specialist company in Air, Land and Sea operations located in South Florida (Also known as Non-destructive Evaluation, NDE or Non-destructive Testing, NDT) . We serve the Miami, Fort Lauderdale, and West Palm Beach areas. NDT Services also performs AOG NDT services within the entire State of Florida, such as Tampa, Orlando, Jacksonville, Tallahassee, etc. We don't stop there, we also travel the entire United States, as well as Internationally to aid in supplying Testing Services globally. We are Non-destructive Testing specialists. We provide solutions to businesses like you that need consultation, testings, or management for your Non-destructive Testing requirements. What does it take to be good at what we do? Passion, Dedication, a lot of Training and a lot of Coffee.

Words and Phrases

quality control 质量控制
non-destructive testing (NDT) 无损检测
non-destructive evaluation (NDE) 无损评价
non-destructive inspection (NDI) 无损检测
non-contact [nɒnˈkɒntækt] adj. 没有接触的
radiographic testing (RT) X 射线检测
ultrasonics inspection (UT) 超声检测

magnetic particle inspection (MT) 磁粉检测
liquid penetrant inspection (PT) 渗透检测
eddy current testing (ET) 涡流检测
thermography [θəˈmɔgrəfi] n. 热成像技术，温度记录法
visual-optical testing 目视检测

NDT Certification

"How Do I Become NDT Certified?"

There are multiple steps to become certified in an NDT method: training, experience and examination. A common misconception is that NDT courses certify a student upon completion, however, taking an NDT course does not certify a student in an NDT method. The course only satisfies the classroom training requirement in the multistage certification process.

If that's so, then how do you become certified?

1. Certification in a nutshell

Certification is typically done by the employer (There is another option, see Third Party Certification in the "Detailed Information about certification" section below).

The following must be completed to become certified:

(1) Classroom Training—The part Hellier does.

(2) Experience—The part you get by working on the job with a certified person in the method for which certification is sought.

(3) Examination—After the classroom training and experience have been accomplished you will be given four different examinations.

2. Detailed information about certification

This section offers more detailed information about the certification process.

There are several steps to become fully certified in an NDT method.

(1) Select an NDT program.

To become NDT Certified, you must first select an NDT program. Most employers have already selected a program so ask your company which one they participate in. The most common programs in North America: NAS 410, SNT-TC-1A, CP 189, ISO 9712, ACCP and CGSB. Each program will outline the various requirements for certification.

(2) Training hours/education.

Just about all NDT programs require training based on topical training course outlines taught by subject-matter experts. Training must cover the Body of Knowledge and fulfill requirements for each level of qualification defined by the NDT program.

To view the amount of Training Hours necessary to become certified, visit the ASNT Certification webpage, select the desired ASNT certification program and view the "Initial Qualification Requirements".

(3) Experience.

ASNT requires "Hours in Method" and "Total Hours in NDT" experience in order to be certified. The "Hours in Method" refers to the amount of experience in a specific method, while "Total hours in NDT" refers to the amount of experience in all NDT methods. The amount of experience required (Method or Total) varies depending on the method and the program you wish to participate in.

To view the amount of Experience necessary to become certified, visit the ASNT Certification webpage, select the desired ASNT certification program and view the "Initial Qualification Requirements".

(4) Certification exam.

Third Party Certification Vs Employer Certification

Employer-based certification : Employers are responsible for NDT training, testing and documentation for employees according to their qualification and certification procedure. The employer will provide a formal certificate or letter authorizing their employee to perform NDT tasks, and certification will expire after a period of time (typically 3 or 5 years) or when an employee leaves the company. An Employee also loses certification if he/she no longer meets the company requirements, for example no current eye exam or not passing a re-certification exam.

Central certification (Third Party) : Qualification examinations are administered by an independent third-party certification body based on a central certification standard. Candidates must provide documentation of training and experience and a certificate will be issued if the examination is successfully completed. The advantage is that employees maintain certification after leaving a company, but each employer has a choice whether to accept third-party certificates as proof of qualification or not.

The exams commonly associated with certification are:

① General—Tests your general knowledge in the theory of the method.

② Specific—Tests your ability to apply the knowledge in procedures, specifications, codes and use of the equipment used by the employer.

③ Practical—Tests your ability to actually use the equipment and perform the test as well as to evaluate test samples for acceptability.

④ Vision—You will be given a vision exam on an annual basis to test your visual acuity.

Item 2
02 Magnetic Partical Testing

Learning Objectives

1. Knowledge objectives

(1) To grasp the words, related terms and abbreviations about MT.

(2) To grasp the classification about MT system.

(3) To know the Instruments and Equipment of MT system.

(4) To know the testing procedure of MT system.

2. Competence objectives

(1) To be able to read and understand frequently used & complex sentence patterns, capitalized English materials and obtain key information quickly.

(2) To be able to communicate with English speakers about the topic freely.

(3) To be able to fill in the job cards in English.

3. Quality objectives

(1) To be able to self-study with the help of aviation dictionaries, the Internet and other resources.

(2) To do a good job of detection of safety protection.

2.1 Theory of Magnetic Partical Testing

Magnetic Field

The simple observations of attracting and repelling indicate that some force field surrounds the magnetized rod. Although invisible, this force field is clearly three-dimensional because the attraction or repulsion can be experienced all around the rod. A two-dimensional slice through this field can be made visible by placing a sheet of plain white paper over a bar magnet and sprinkling ferromagnetic particles onto it. The particles will collect around the lines of force in the magnetic field, producing an image such as the one shown in Fig. 2-1.

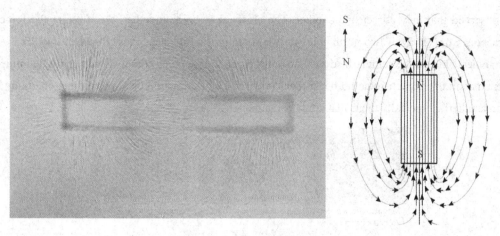

Fig. 2-1 Magnetic Field

This image is called a "magnetograph" and the lines of force are referred to as lines of "magnetic flux". Lines of magnetic flux will flow in unbroken paths that do not cross each other. Each line of force forms a closed loop that flows through and around the magnet. The word "flow" suggests some sort of direction of movement and by convention this direction is said to be the direction that would be taken by a "unit north pole" placed at the north pole of a bar magnet. A unit north pole is an imaginary concept of a particle with no corresponding south pole. Such a particle would be repelled by the magnet's north pole and attracted to the south pole. In other words, magnetic flow is from its north pole to its south pole through the air around the magnet and in order to complete the magnetic circuit, flow will be from the south pole to the north pole within the magnet.

Polarity (Fig. 2-2)

Many of the basic principles of magnetism can be deduced by simple observation of the behavior of a magnetized rod and its interaction with ferromagnetic materials, including other magnetized rods. If the rod is suspended at its center, it will eventually align itself with the Earth's magnetic field so that one end points to geographic north and the other end to the south. If the north-pointing end is identified, it will be found that it is always this end that points north. By convention, this end of the rod is called the "northseeking pole", usually abbreviated as "north pole" and the other end is called the "south pole".

Fig. 2-2 Polarity

Magnetic Forces (Fig. 2-3)

When the north pole of one magnetized rod is placed close to the south pole of another, it will be observed that they attract one another. The closer they come together, the stronger the force of attraction. Conversely, if two north poles or two south poles are placed close together, they will repel each other. This can be summarized as "like poles repel, unlike poles attract". One way of defining the phenomenon of magnetism could be "a mechanical force of attraction or repulsion that one body has upon another", especially those that are ferromagnetic.

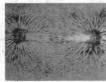

Magnetic lines of force around a bar magnet Opposite poles attracting Similar poles repelling

Fig. 2-3 Magnetic Forces

Flux Density

The flowing force of magnetism is called "magnetic flux". The magnetograph image does not show the direction of flux flow, but it can be seen from the magnetograph that the area of maximum flux concentration (flux density) is at the poles. Flux density is defined as "the number of lines of force per unit area". The unit area referred to is a slice taken perpendicular to the lines of force. Flux density is measured in Gauss or Tesla, the Tesla being the current unit, and flux density is given the symbol "β" (beta).

Magnetizing Force

The total number of lines of force making up a magnetic field determines the strength of the force of attraction or repulsion that can be exerted by the magnet and is known as the "magnetizing force" and given the symbol "H". The unit of magnetizing force are shown in Table 2-1.

Table 2-1 Unit of Magnetizing Force

Magnetic Field Type	SI	cgs	U.S. Customary
Magnetic Flux φ	Webber	Maxwell	Lines (or kilo-lines)
Magnetic Field Intensity H	Amp Turns per meter	Oerstead	Amp turns per inch
Magnetic Flux Density, B	Tesla	Gauss	Lines per square inch (or kilo-lines per square inch)

$$1 \text{ Webber} = 10^8 \text{ Maxwell} = 10^5 \text{ kilo-line}$$

$$1 \frac{\text{Amp Turn}}{\text{m}} = 1.257 \times 10^{-2} \text{ Oerstead} = 2.540 \times 10^{-2} \frac{\text{Amp Turn}}{\text{in.}} \text{①}$$

$$1 \text{ Tesla} = 10^4 \text{ Gauss} = 64.52 \frac{\text{kilo-line}}{\text{in.}^2}$$

〖Point 1〗 Leakage magnetic field (Fig. 2-4)

After the ferromagnetic material workpiece is magnetized, due to the existence of discontinuity,

---

① 1in.≈2.54.cm.

the magnetic field of the workpiece surface and near surface is distorted locally, and the magnetic powder applied to the workpiece surface is adsorbed to form visible magnetic marks under suitable illumination, thus showing the position, size, shape and severity of the discontinuity. Also known as magnetic particle testing is one of the five conventional methods of non-destructive testing.

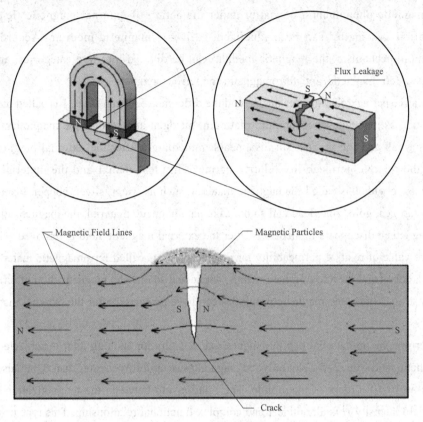

Fig. 2-4 Leakage Magnetic Field

The factors affecting the magnetic field leakage are as follows:

(1) The intensity of the applied magnetic field must be greater than that corresponding to the maximum permeability. The permeability decreases, the magnetoresistance increases and the leakage field increases.

(2) The shallower the defect is buried, the more perpendicular the defect is to the surface, the greater the depth ratio of the defect is, the greater the magnetic flux leakage field is.

(3) The same defect is affected by the surface covering of the workpiece. The thinner the surface covering of the workpiece, the greater the magnetic flux leakage field.

(4) The influence of the material and state of the workpiece on the grain size of the workpiece itself, the amount of carbon content, heat treatment and cold processing will affect the magnetic flux leakage field.

Magnetic powder detection is a method of observing defects with magnetic powder as display medium. According to the type of magnetic powder medium applied during magnetization, the detection method is divided into wet method and dry method, and according to the time of applying magnetic powder on the workpiece, the test method is divided into continuous method and remanent method.

■ 〖Point 2〗 Ferro-Magnetic Material

The magnetic phenomenon of matter under the action of external magnetic field is called magnetization. All matter can be magnetized, so it is a magnetic medium. According to the magnetization mechanism, the magnetic medium can be divided into five categories: anti-magnet, paramagnetic, ferromagnetic, anti-ferromagnetic and ferrous magnets.

The relative permeability greater than 1, and the difference between 1 and 1 is called paramagnetic material, such as manganese, chromium, platinum, nitrogen and so on. The magnetization of this material is weak and the magnetism disappears immediately after the external magnetic field is removed; the relative permeability relative permeability less than 1 and the material with very slight difference with 1 is called diamagnetic material, such as water, silver, copper, bismuth, sulfur, chlorine, hydrogen, gold, zinc, lead and so on. Like paramagnetic material, the magnetization is weak and the magnetism disappears immediately after the external magnetic field is removed; The material with a large value of relative permeability far greater than 1 is called ferromagnetic material, such as iron, nickel, cobalt and the alloy of these metals, ferrite and so on. The magnetization of ferromagnetic material is strong, and some magnetic properties can have preserved after the external magnetic field is removed.

The ferromagnetic material can be magnetized to saturation as long as it is subjected to a very small magnetic field (Fig. 2-5). Not only the magnetic susceptibility greater than 0, but also the value is as large as $10\text{-}10^6$ orders of magnitude. The relationship between the magnetization M and the magnetic field intensity H is a nonlinear and complex functional relationship. This type of magnetism is called ferromagnetism.

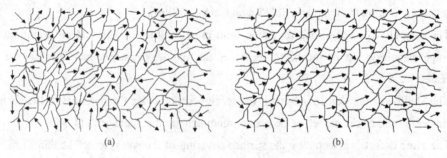

Fig. 2-5 Domains

(a) Iron Unmagnetized—Domains Magnetized in Random Direction;

(b) Iron Strongly Magnetized—Domains Magnetized in Roughly the Same Direction

Ferromagnetic material has ferromagnetism only below Curie temperature and above Curie temperature, the directional arrangement of atomic magnetic moment is destroyed because of the interference of crystal thermal motion, which makes ferromagnetism disappear and then the changes to paramagnetism (Fig. 2-6).

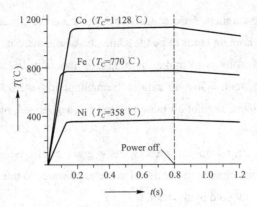

Fig. 2-6 Temperature/Time Profiles and Curie-point Temperatures

Ferromagnetic materials have the following three basic properties. First, the permeability is not constant and decreases with increasing magnetic induction intensity. The second characteristic, the magnetic induction intensity has a saturation value. Third, there is remanence and hysteresis loss in repeated magnetization.

Ferromagnetic materials can be divided into the following three categories. Category I, soft magnetic materials. This kind of material has the characteristics of large permeability and small hysteresis loss. Its hysteresis loop is narrow and steep, the coercive magnetic force is small, it is easy to magnetize, but it is also easy to demagnetize, so it is suitable for repeated magnetization. Common soft magnetic materials are cast steel and silicon steel sheet and so on. Silicon steel sheets of various specifications with a thickness of 0.35-1 mm are important magnetic conductivity materials for the manufacture of transformers, motors and ac electromagnets. the saturation value of magnetic induction strength of hot rolled silicon steel sheets can reach 14 000 Gauss, and that of cold rolled silicon steel sheets can reach 18 000 Gauss. Cast iron can be used to make the housing and parts of the motor, its saturation value does not exceed 10 000 Gauss.

Class II, hard magnetic materials. This kind of material is characterized by large remanence, large coercive magnetic force, wide hysteresis loop fat and wide, difficult magnetization and need to use very strong external magnetic field. The variety of hard magnetic materials is very diverse and is constantly being studied and improved. More widely used are alloys containing different components of aluminum, nickel, iron, such as aluminum 11%, nickel 22%, the rest of the alloys are iron, with a residual magnetic value of 7 000 Gauss. Using barium ferrite, strontium-calcium ferrite and other materials, a new type of hard magnetic ferrite can be made. Its coercive magnetic force is higher than that of alloy hard magnetic material, its specific gravity is small, the price is low, but the remanence is low. Hard magnetic materials can be used to make permanent magnets for various purposes.

■ 〖Point 3〗 Magnetizing

Magnetization is the process of magnetizing substances that are not magnetic. Magnetic materials are divided into many tiny regions, each of which is called a domain, and each domain has its own magnetic moment (that is a tiny magnetic field) . In general, the magnetic moment direction of each domain is different and the magnetic field cancels each other, so the whole material is not magnetic. When the direction of each domain tends to be the same, the whole material shows magnetism.

The so-called magnetization is to make the magnetic moment direction of magnetic domains in magnetic materials consistent. When an external nonmagnetic material is put into another strong magnetic field, it is magnetized, but not all materials can be magnetized, only a few metals and metal compounds can be magnetized.

The selection of magnetization parameters has a great relationship with the magnetization specification—standard magnetization specification, relaxed magnetization specification and strict magnetization specification adopted by the user.

Standard magnetization specification can clearly display all defects on the workpiece. For example: cracks with a depth of more than 0.05 mm, smaller hair lines and non-metallic inclusions. Generally used for Testing of higher-required workpieces.

Relaxation of magnetization specification: can clearly show the workpiece of various cracks and larger defects. Suitable for Testing of ordinary workpiece.

■ 〖Point 4〗 Relationship between Magnetic Field Direction and Defect Detection of Magnetic Powder（Fig. 2-7）

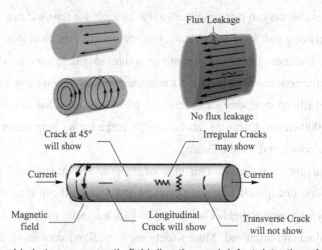

Fig. 2-7　Relationship between magnetic field direction and defect detection of magnetic powder

The ability of magnetic powder detection depends on the size of the applied magnetic field and the extension direction of the defect, and is also related to the position, size and shape of the defect. When the magnetic field direction is perpendicular to the defect extension direction, the magnetic leakage field at the defect is large and the detection sensitivity is high: when the angle between the magnetic field direction and the defect extension direction is 45°, the defect can be displayed, but the sensitivity is reduced. Because the defects in the workpiece have various orientations and are difficult to predict, it should be based on the geometry of the workpiece. Different methods are used for circumferential, longitudinal or multidirectional magnetization of the workpiece directly, indirectly or through inductive current in order to make the magnetic field direction perpendicular to the possible defects of the workpiece as far as possible. Multiple magnetization methods can be used in combination with workpiece size, structure and shape to detect defects in all directions (Fig. 2-8).

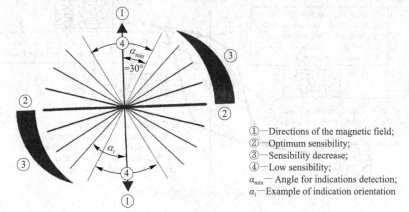

Fig. 2-8 Directions of Detectable Indications

①—Directions of the magnetic field;
②—Optimum sensibility;
③—Sensibility decrease;
④—Low sensibility;
α_{min}— Angle for indications detection;
α_i—Example of indication orientation

In addition, the factors to be considered in the selection of magnetization method are: the size of the workpiece; the shape and structure of the workpiece; the surface state of the workpiece. According to the past fracture of the workpiece and the stress distribution of each part, the location and direction of the possible defects are analyzed and the appropriate magnetization method is selected.

〖Point 5〗 Magnetic Saturation Magnetic Hysteres is Loop

The lines of force in a magnetic field repel adjacent lines flowing in the same direction. As the flux density increases, the force of repulsion increases. For a given material there is a maximum value for flux density that can be sustained. Upon reaching this value, the material is said to be "saturated". As the flux density is increased towards saturation, the reluctance of the material increases and the permeability decreases towards that of a vacuum. At saturation, any further increase in magnetizing force finds that the path of least reluctance is now through the air surrounding the material and the excess flux extends out into the air.

Hysteresis

When a ferromagnetic material is magnetized in one direction, it will not relax back to zero magnetization when the imposed magnetizing field is removed. It must be driven back to zero by a field in the opposite direction. If an alternating magnetic field is applied to the material, its magnetization will trace out a loop called a hysteresis loop (Fig. 2-9). The lack of traceability of the magnetization curve is the property called hysteresis and it is related to the existence of magnetic domains in the material. Once the magnetic domains are reoriented, it takes some energy to turn them back again. This property of ferromagnetic materials is useful as a magnetic "memory". Some compositions of ferromagnetic materials will retain an imposed magnetization indefinitely and are useful as "permanent magnets". The magnetic memory aspects of iron and chromium oxides make them useful in audio tape recording and for the magnetic storage of data on computer disks.

It is customary to plot the magnetization M of the sample as a function of the magnetic field strength H, since H is a measure of the externally applied field which drives the magnetization .

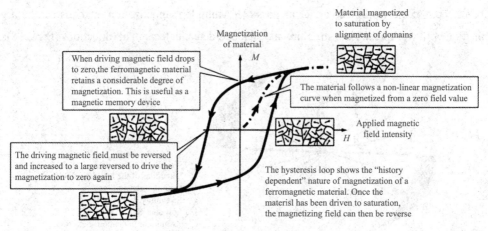

Fig. 2-9 Hysteresis Loop

Because of hysteresis (Fig. 2-10), an input signal at the level indicated by the dashed line could give a magnetization anywhere between C and D, depending upon the immediate previous history of the tape (i.e., the signal which preceded it). This clearly unacceptable situation is remedied by the bias signal which cycles the oxide grains around their hysteresis loops so quickly that the magnetization averages to zero when no signal is applied. The result of the bias signal is like a magnetic eddy which settles down to zero if there is no signal superimposed upon it. If there is a signal, it offsets the bias signal so that it leaves a remnant magnetization proportional to the signal offset.

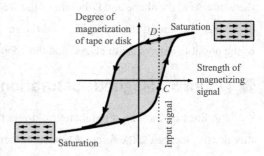

Fig. 2-10 Hysteresis

There is considerable variation in the hysteresis of different magnetic materials (Fig. 2-11).

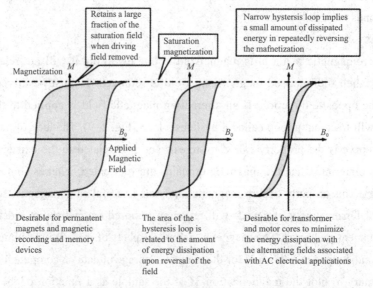

Fig. 2-11 The Hysteresis of Different Magnetic Materials

〖Point 6〗 History of Magnetic Particle Inspection

Magnetism is the ability of matter to attract other matter to itself. The ancient Greeks were the first to discover this phenomenon in a mineral they named magnetite. Later on Bergmann, Becquerel, and Faraday discovered that all matter including liquids and gasses were affected by magnetism, but only a few responded to a noticeable extent.

The earliest known use of magnetism to inspect an object took place as early as 1868. Cannon barrels were checked for defects by magnetizing the barrel then sliding a magnetic compass along the barrel's length. These early inspectors were able to locate flaws in the barrels by monitoring the needle of the compass. This was a form of non-destructive testing but the term was not commonly used until some time after World War I.

In the early 1920s, William Hoke realized that magnetic particles (colored metal shavings) could be used with magnetism as a means of locating defects. Hoke discovered that a surface or subsurface flaw in a magnetized material caused the magnetic field to distort and extend beyond the part. This discovery was brought to his attention in the machine shop. He noticed that the metallic grindings from hard steel parts (held by a magnetic chuck while being ground) formed patterns on the face of the parts which corresponded to the cracks in the surface. Applying a fine ferromagnetic powder to the parts caused a build up of powder over flaws and formed a visible indication. The image shows a 1928 Electyro-Magnetic Steel Testing Device (MPI) made by the Equipment and Engineering Company Ltd. (ECO) of Strand, England.

In the early 1930s, magnetic particle inspection was quickly replacing the oil-and-whiting method (an early form of the liquid penetrant inspection) as the method of choice by the railroad industry to inspect steam engine boilers, wheels, axles and tracks. Today, the MPI inspection method is used extensively to check for flaws in a large variety of manufactured materials and components. MPI is used to check materials such as steel bar stock for seams and other flaws prior to investing machining time during the manufacturing of a component. Critical automotive components are inspected for flaws after fabrication to ensure that defective parts are not placed into service. MPI is used to inspect some highly loaded components that have been in-service for a period of time. For example, many components of high performance racecars are inspected whenever the engine, drive train or another system undergoes an overhaul. MPI is also used to evaluate the integrity of structural welds on bridges, storage tanks and other safety critical structures.

Put into Practice

1. Translate the English terms in the Fig. 2-12.

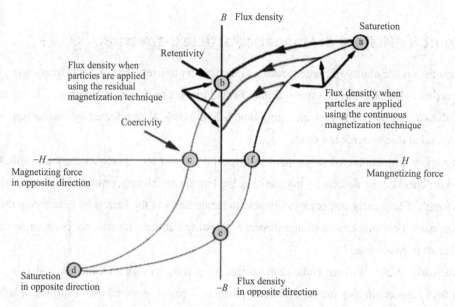

Fig. 2-12 The Hysteresis of Magnetic Materials

2. It is pointed out that the magnetization method in Fig.2-13 belongs to the weekly magnetization and is still longitudinal magnetization.

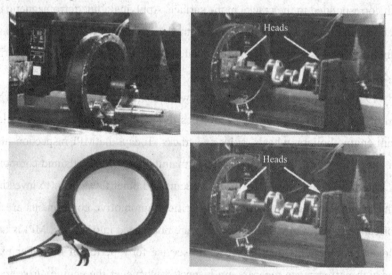

Fig. 2-13 Magnetization Method

3. Draw the direction of the defect that can be found by magnetizing in Fig. 2-14.

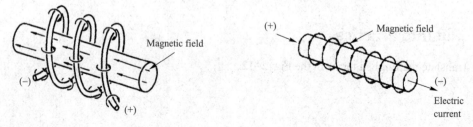

Fig. 2-14 The Direction of the Defect

2.2 Classification of Magnetic Particle Testing

■〖Point 1〗 Classification of Method

Classification of methods:

According to the magnetization direction of the workpiece, it can be divided into circumferential magnetization method, longitudinal magnetization method, composite magnetization method and rotary magnetization method.

It can be divided into DC magnetization, half wave DC magnetization and AC magnetization according to the different magnetization current.

It can be divided into dry powder method and wet powder method according to the preparation of magnetic powder used in flaw detection.

It can be divided into continuous method and remanent method according to the time of applying magnetic powder on the workpiece.

■〖Point 2〗 Dry Powder Method and Wet Powder Method (Fig. 2-15)

Magnetic powder suspended in oil, water or other liquid media is used called wet method. It is used to distribute the magnetic suspension evenly on the surface of the workpiece during the detection process. The shape and size of the defect are displayed by using the flow of the carrier liquid and the attraction of the magnetic leakage field to the magnetic powder. In wet detection, the magnetic powder particles can be used because of the dispersion and suspension performance of magnetic suspension. Therefore, it has high detection sensitivity. It is especially suitable for detecting small defects on the surface, such as fatigue cracks, grinding cracks, etc. Wet methods are often used in conjunction with stationary equipment and with mobile and portable equipment. Magnetic suspensions for wet processes can be recycled.

Dry method is called dry powder method. In some special cases, when wet method can not be used to detect, the special dry magnetic powder is applied directly to the magnetized workpiece according to the procedure, and the defect of the workpiece shows magnetic mark. Dry testing is mostly used for local area Testing of large casting, forging blank and large structural parts and

Fig. 2-15 Dry Powder Method and Wet Powder Method

welding parts, usually in conjunction with portable equipment.

■ 〖Point 3〗 Continuous and Remanence Methods

Continuous method, also called accessory magnetic field method or present magnetic method, is to apply magnetic powder or magnetic suspension to the workpiece for magnetic particle testing under the action of external magnetic field. The observation and evaluation of the workpiece can be carried out under the action of external magnetic field or after interrupting the magnetic field.

The remanence method is to magnetize the workpiece first, then pour the suspension on the workpiece and observe the magnetic powder after gathering. This is the use of material residual magnetic detection method, so called remnant magnetic method.

■ 〖Point 4〗 DC Magnetization, AC Magnetization (Fig. 2-16)

At present, pure DC is rarely used, while single-phase half-wave or full-wave, three-phase half-wave or full-wave rectifier is used.

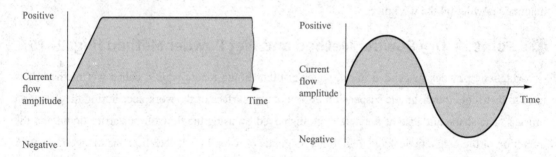

Fig. 2-16 DC and AC

AC of the general frequency, AC magnetization, the peak current can work, but the current value usually represents the effective value. Both AC magnetization and DC magnetization can detect surface and near surface defects. AC magnetization has high sensitivity for surface defect detection, and DC magnetization has strong ability for surface defect detection. In general, the larger the pulsating component contained in the current, the weaker the ability to detect internal defects. DC magnetization is used for remanence stability and AC magnetization is not stable enough. The power off phase controller can solve this problem. DC magnetization demagnetization is difficult; AC magnetization demagnetization is easy. At present, AC magnetization is generally recommended, and continuous method is used and phase control is generally added at this time. AC continuous magnetization, simple structure, low cost, magnetization effect can generally meet the requirements. There is also the use of AC and DC mixed magnetization, at this time AC magnetization in front, DC magnetization in the back, easy demagnetization.

Rectified Alternating Current (Fig. 2-17)

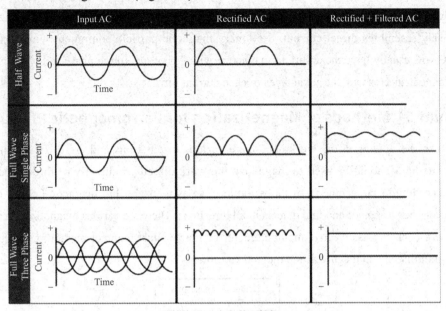

Fig. 2-17 Rectified Alternating Current

Clearly, the skin effect limits the use of AC since many inspection applications call for the detection of subsurface defects. However, the convenient access to AC, drives its use beyond surface flaw inspections. Luckily, AC can be converted to current that is very much like DC through the process of rectification. With the use of rectifiers, the reversing AC can be converted to a one directional current. The three commonly used types of rectified current are described below.

Half Wave Rectified Alternating Current (HWAC)

When single phase alternating current is passed through a rectifier, current is allowed to flow in only one direction. The reverse half of each cycle is blocked out so that a one directional, pulsating current is produced. The current rises from zero to a maximum and then returns to zero. No current flows during the time when the reverse cycle is blocked out. The HWAC repeats at same rate as the unrectified current (60 hertz typical). Since half of the current is blocked out, the amperage is half of the unaltered AC.

This type of current is often referred to as half wave DC or pulsating DC. The pulsation of the HWAC helps magnetic particle indications form by vibrating the particles and giving them added mobility. This added mobility is especially important when using dry particles. The pulsation is reported to significantly improve inspection sensitivity. HWAC is most often used to power electromagnetic yokes.

Full Wave Rectified Alternating Current (FWAC) (Single Phase)

Full wave rectification inverts the negative current to positive current rather than blocking it out. This produces a pulsating DC with no interval between the pulses. Filtering is usually performed to soften the sharp polarity switching in the rectified current. While particle mobility is not as good as half-wave AC due to the reduction in pulsation, the depth of the subsurface magnetic field is improved.

Three Phase Full Wave Rectified Alternating Current

Three phase current is often used to power industrial equipment because it has more favorable

41

power transmission and line loading characteristics. This type of electrical current is also highly desirable for magnetic particle testing because when it is rectified and filtered, the resulting current very closely resembles direct current. Stationary magnetic particle equipment wired with three phase AC will usually have the ability to magnetize with AC or DC (three phase full wave rectified), providing the inspector with the advantages of each current form.

〖Point 5〗 Methods of Magnetization for Ferromagnetic Products

Because the defects in the workpiece have various orientations and are difficult to predict, different methods should be used to magnetize the work directly, indirectly or through inductive current according to the geometry of the workpiece, so as to establish a magnetic field in different directions on the workpiece and find defects in all directions. There are various magnetization methods (Fig. 2-18), such as pass electric method, center conductor dust, contact method, coil dust, yoke method, multidirectional magnetization method and so on.

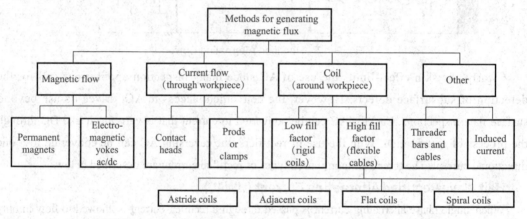

NOTE: For complicated shapes, the use of two or more of these methods may be requirde.

Fig. 2-18   Methods of Magnetzation

1. Circumferential magnetization

The magnetization method of the workpiece directly electrified or allowing the current to pass through a conductor placed parallel to the axial direction of the workpiece is called circumferential magnetization. The aim is to establish a closed circumferential magnetic field around the workpiece and perpendicular to the axial direction of the workpiece to find defects whose orientation is basically parallel to the current direction (that is axial defects).

For small parts, the whole circumferential magnetization (Fig.2-19) of the tested parts is made by direct electrification or central conductor electrification. The contact method (direct electrification) and parallel cable method (auxiliary electrification) are used to detect the magnetic powder of large structure.

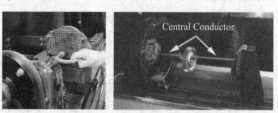

Fig. 2-19   Circumferential Magnetization

(1) Contact method. The method of introducing magnetization current into the workpiece for local magnetization by two electrode contacts is called contact method (Fig. 2–20). In order to avoid missing defects, the same part should be detected perpendicular to each other at least twice by changing the orientation of the contact line.

Portable Prod Unit

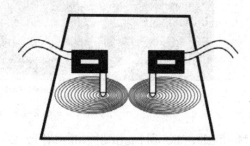

Fig. 2-20 Prode

(2) Parallel cable method. The longitudinal cracks in the area can be detected by circumferential magnetization of cables parallel to the detected area.

2. Longitudinal magnetization

The purpose of longitudinal magnetization is to establish a longitudinal magnetic field distributed along its axial direction in the workpiece using an excitation coil surrounding the workpiece or yoke core to find defects whose orientation is basically perpendicular to the axial direction of the workpiece (circumferential or radial defects).

The commonly used methods are yoke method and coil method.

(1) Yoke method (Fig. 2–21). The whole or local longitudinal magnetization can be made by contacting the two poles of the yoke or permanent yoke with the inspected workpiece. If the two sections of the work can be clamped between the poles of the yoke to form a closed magnetic path, the workpiece can be magnetized vertically, otherwise it is local magnetization. When local magnetization is made, the magnetic force line between the two poles of the yoke is roughly parallel to the connecting line of the two poles, and the defects whose orientation is basically perpendicular to the connecting line of the two poles can be detected.

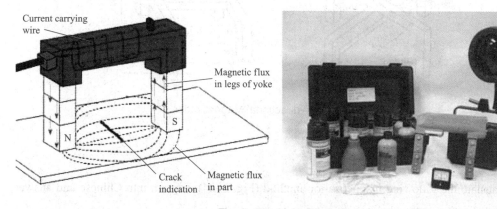

Fig. 2-21 Yoke Method

(2) Coil method. The method of longitudinal magnetization of the tested workpiece by spiral coil (Fig. 2-22) is called coil method. The coil method can be used to detect the magnetic powder of the pipe ring weld, and the longitudinal crack of the ring weld and its heat affected zone can be found.

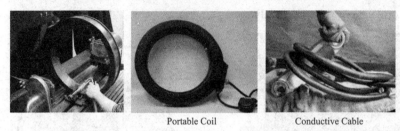

Fig. 2-22 Coil

3. Multi-directional magnetization (Fig. 2-23)

Allow the component to be magnetized in two directions, longitudinally and circumferentially, in rapid succession. Therefore, inspections are conducted without the need for a second shot. In multidirectional units, the two fields are balanced so that the field strengths are equal in both directions. These quickly changing balanced fields produce a multidirectional field in the component providing detection of defects lying in more than one direction.

Just as in conventional wet-horizontal systems, the electrical current used in multidirectional magnetization may be alternating, half-wave direct or full-wave. It is also possible to use a combination of currents depending on the test applications. Multidirectional magnetization can be used for a large number of production applications and high volume inspections.

To determine adequate field strength and balance of the rapidly changing fields, technique development requires a little more effort when multidirectional equipment is used. It is desirable to develop the technique using a component with known defects oriented in at least two directions or a manufactured defect standard. Quantitative Quality Indicators (QQI) are also often used to verify the strength and direction of magnetic fields.

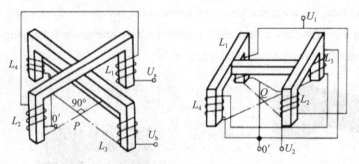

Fig. 2-23 Multi-directional Magnetization

Put into Practice

1. Translate the following magnetization method (Fig. 2-24) diagram into Chinese and answer which magnetization method belongs to.

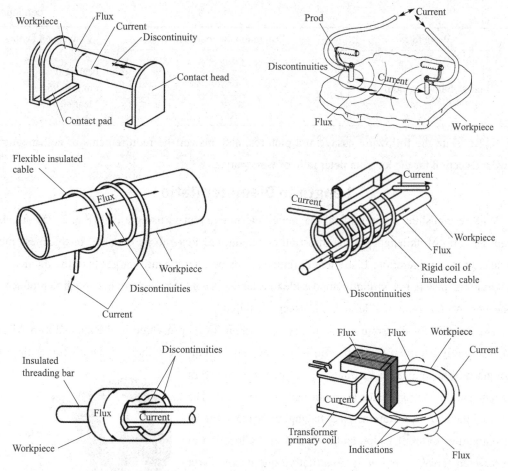

Fig.2-24 Magnetization Method

2. Answer the type of magnetized current, current peak, average and effective value in Table 2-2.

Table 2-2 The Type of Magnetized Current

Wave form	Diagrammatic representation	Conversion factors
Aternating current (ac)		Peak=1.0 r.m.s.=0.707 Mean=0.636
Direct current (dc)		Peak=1.0 r.m.s.=1.0 Mean=1.0
Half wave rectified ac (HWR)		Peak=1.0 r.m.s.=0.5 Mean=0.318
Full wave rectified ac (FWR)		Peak=1.0 r.m.s.=0.707 Mean=0.636

Wave form	Diagrammatic representation	Conversion factors
3-phase full wave rectified ac		Peak=1.0 r.m.s.=0.95 Mean=0.955

3. Translate the following essays and pictures, and answer the requirements of coil magnetic powder detection for the long diameter ratio of work parts?

Length to Diameter Ratio

When establishing a longitudinal magnetic field in component using a coil (Fig. 2–25) or cable wrap, the ratio of its length (in the direction of the desired field) to its diameter or thickness must be taken into consideration. If the length dimension is not significantly larger than the diameter or thickness dimension, it is virtually impossible to establish a field strength strong enough to produce an indication. An L/D ratio of at least two is usually required.

The formula for determining the necessary current levels presented in the appendix of ASTM 1 444 are only useful if the L/D ratio is greater than two and less than 15. Don't forget that the formula only provides an estimate of the necessary current strength and this strength must be confirmed in other ways. The preferred method is to examine parts having known or artificial discontinuities of similar type and size, and in the location of the targeted flaws; or by using quantitative quality indicator (notched) shims. A second method is to use gaussmeter with a tangential field hall effect probe to measure the field strength, which must be in the range of 30 to 60 G.

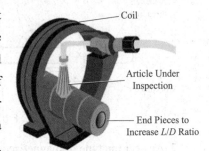

Fig.2–25 Coil

2.3 Instruments and Equipment of Magnetic Particle Testing

■ 〖Point 1〗 Classification of Magnetic Particle Testing Instruments

The classification of magnetic powder detection equipment can be divided into fixed, mobile and portable according to the weight and mobility of the equipment. According to the combination of equipment can be divided into two types: integrated type and discrete type. The integrated magnetic powder flaw detector is a separate device, which combines the magnetized power supply, solenoid coil, workpiece clamping device, magnetic suspension spraying device, lighting device and demagnetization device according to its function and is combined into a system flaw detector during flaw detection. Fixed flaw detector belongs to one type, easy to use and operate. Mobile and portable flaw detectors are discrete and easy to move and combine in the field.

1. Fixed magnetic particle detector (Fig. 2-26)

The fixed magnetic particle detector is large in volume and weight, and the rated circumferential magnetization current is generally from 1 000 to 10 000 A. The utility model can carry out electrification method, central conductor method, inductive current method, coil method, magnetic rail method, integral magnetization or composite magnetization, etc., with lighting device, demagnetization device and magnetic suspension stirring, spraying device, magnetization chuck of clamping workpiece and worktable and grille placed to workpiece, which is suitable for flaw detection of small and medium-sized workpiece. Contacts and cables are also often available. In order to carry on the testing to the worktable has the difficulty large workpiece.

Fig. 2-26 Fixed Magnetic Particle Detector

2. Mobile magnetic particle detector (Fig. 2-27)

The rated circumferential magnetization current of mobile magnetic particle flaw detector generally ranges from 500 to 800 A. The main body is magnetized power supply, which can provide the magnetization current of AC and unidirectional half wave rectifier. The accessories include contact, clamp, open and close magnetization coil and soft cable. Such equipment is generally equipped with rollers to push, or hoisting in the car to the testing site. Testing of large workpieces.

Fig. 2-27 Mobile Magnetic Particle Detector

3. Portable flaw detector (Fig. 2-28)

The portable flaw detector has the characteristics of small size, light weight and easy to carry and the rated circumferential magnetization current generally from 500 to 2 000 A. Suitable for field, high altitude and field testing, generally used to test boiler pressure vessel and pressure pipe welding and in situ testing of aircraft, trains, ships or local testing of large workpieces. The commonly used instruments include small magnetic particle flaw detector with contact, electromagnetic rail, cross magnetic rail or permanent magnet. The instrument handle is equipped with a micro current switch, control pain, power off and brake attenuation demagnetization.

Fig. 2-28 Portable Flaw Detector

Foreign countries attach great importance to the development of magnetic powder detection equipment, because only the progress of testing equipment can bring successful application to magnetic powder detection. Now foreign magnetic particle testing equipment from fixed magnetic particle testing machine, mobile magnetic particle testing machine, to portable magnetic particle testing machine, from semi-automatic magnetic particle testing machine, automatic magnetic particle testing machine to special magnetic particle testing equipment, from unidirectional magnetization to multi-directional magnetization, the equipment has been serialized and commercialized. Because thyristor and other electronic components are used in magnetic powder detection equipment, intelligent equipment has emerged in large numbers, which can preset magnetization specifications and reasonable process parameters. The detection and automatic operation of fluorescent magnetic powder has also been successfully carried out in foreign countries, and the fluorescent magnetic powder scanning and laser flying point scanning system of the imperial TV photodetector has realized the automation of the observation stage of magnetic powder detection. The sensitivity and reliability of the detection are greatly improved. It represents the new achievement of contemporary magnetic powder detection.

In recent years, magnetic powder detection equipment in China has developed rapidly, and has been serialized. The performance of three all-wave DC flaw detection ultra-low frequency demagnetization equipment has reached the level of similar equipment abroad. The AC flaw detection machine is used for remanence test with a power-off phase controller to ensure remanence stability. It is the characteristic of our country. By using SCR technology, the power-off phase controller can replace the stepless regulation of magnetization current of self-root transformer. Intelligent equipment has been produced and applied. A magnetic particle detector for photoelectric scanning image recognition has been developed successfully. The experimental research on the reality of computer processing magnetic marks has made great progress, and there are many applications in automation and semi-automation equipment.

■ 〖Point 2〗 Structure and Composition of Magnetic Powder Testing Equipment

Magnetic particle testing equipment (Fig. 2-29) is generally composed of magnetization device, electrical control, workpiece rotation device, magnetic suspension spray recovery, hydraulic or pneumatic control system and so on.

The workpiece rotation device is mainly used to support and drive the workpiece rotation, so that the workpiece can be magnetized stably and uniformly, so that the magnetic powder or magnetic suspension can be sprayed evenly on the workpiece and the magnetic mark of the workpiece can be easily observed and found at the same time.

Magnetic particle testing equipment is generally composed of magnetization

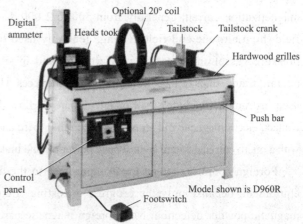

Fig. 2-29 Wet Horizional Unit

device, electrical control, workpiece rotation device, magnetic suspension spray recovery, hydraulic or pneumatic control system and so on.

The workpiece rotation device is mainly used to support and drive the workpiece rotation, so that the workpiece can be magnetized stably and uniformly, so that the magnetic powder or magnetic suspension can be sprayed evenly on the workpiece and the magnetic mark of the workpiece can be easily observed and found at the same time.

Magnetic particle testing equipment is generally composed of magnetization device, electrical control (Fig. 2–30), workpiece rotation device, magnetic suspension spray recovery, hydraulic or pneumatic control system and so on.

Fig. 2–30 Electrical Control

The workpiece rotation device is mainly used to support and drive the workpiece rotation, so that the workpiece can be magnetized stably and uniformly, so that the magnetic powder or magnetic suspension can be sprayed evenly on the workpiece, and the magnetic mark of the workpiece can be easily observed and found at the same time.

So magnetic particle flaw detector is generally equipped with demagnetization device. The demagnetization device is used to recede the remanence in the magnetized specimen. It is generally composed of demagnetizing power supply, demagnetizing coil, controller and so on. Slowly passing the part through the AC coil allows it to demagnetize. In addition, some magnetic particle flaw detectors do not have external demagnetization devices, but use the gradual reduction of direct demagnetization through the alternating current of the parts according to the type of demagnetization current. Fork can be divided into DC demagnetizing device, AC demagnetizing device, ultra low frequency demagnetizing device and so on.

■ 〖Point 3〗 Quantitative Quality Indicator Test Piece Shims QQI

QQIs are artificial flaw (notched) shims that are attached to example parts, commonly used to demonstrate both field strength and direction within a part. Available in several different configurations, QQIs are thin steel shims with etched patterns in circular and cross shapes to provide indications in all directions. The steel alloy and notch dimensions, as specified in AS 5371, are designed to provide indications when the base part is magnetized to at least 30 gausses. The thin shims can conform to curved part surfaces, and they are typically affixed to a part using permanent adhesives.

Remove the corrosion-resistant film from the QQI before use. Solvents such as SKC-S, acetone, or an adhesive remover (Goof Off or equivalent) are recommended. Use caution when handling to prevent damage or distortion.

Instructions

QQIs (Fig. 2-31) must be placed in intimate contact with the notches facing inward towards the part surface. Permanent adhesive such as cyanoacrylate (super glue or equivalent) is recommended. Attach the QQI conforming to the part surface with no gaps or loose areas. No adhesive should remain on the outer Testing surface after the QQI is attached.

QQIs are useful for setting up the magnetization parameters for part-specific techniques, and can be used to create an example part for the daily system performance check of the magnetizing equipment.

They are very important for the setup and balance of multi-directional fields, since they have circular flaws that show indications in all directions simultaneously.

Cellophane tape (Scotch Brand 191 471 or 600 series) may be used to attach QQIs. Care should be taken to only cover the edges of the QQI so that the central notched area is clear. If the tape becomes loose, completely remove the QQI, clean and reattach with new tape.

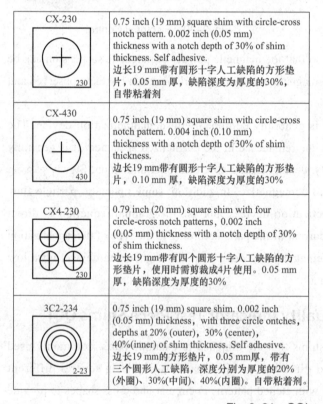

Specifications 符合规范
- AS 5371
- ASME BPVC Section V Article 7
- ASTM E709
- ASTM E1444
- ASTM E3024

Part Number 件号
625551	CX-230, Set of 5 一套五片
625552	CX4-230, Set of 5 一套五片
625553	CX-430, Set of 5 一套五片
625554	3C2-234, Set of 5 一套五片

Fig. 2-31 QQIs

【Point 4】 Mag Particle Test Piece

The Tool Steel Ring is a standardized test piece commonly used with wet bench magnetic particle equipment. The ring is machined from AISI O1 Tool Steel, annealed, tested and certified to meet AS 5282 specifications. Typically used with a 1-inch central conductor, the Tool Steel Ring (Fig. 2-32) has 12 machined holes at increasing depths from the edge and is used to verify the performance of

HWDC, FWDC and 3-phase FWDC magnetizing equipment. Suitable for use with wet or dry materials and visible or fluorescent particles. The number of indications required depends on the waveform and magnetizing current amperage (refer to AS 5282 or ASTM E1444 for more information).

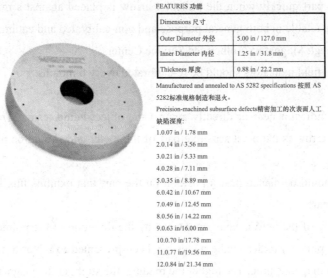

FEATURES 功能

Dimensions 尺寸	
Outer Diameter 外径	5.00 in / 127.0 mm
Inner Diameter 内径	1.25 in / 31.8 mm
Thickness 厚度	0.88 in / 22.2 mm

Manufactured and annealed to AS 5282 specifications 按照 AS 5282标准规格制造和退火。
Precision-machined subsurface defects 精密加工的次表面人工缺陷深度：
1. 0.07 in / 1.78 mm
2. 0.14 in / 3.56 mm
3. 0.21 in / 5.33 mm
4. 0.28 in / 7.11 mm
5. 0.35 in / 8.89 mm
6. 0.42 in / 10.67 mm
7. 0.49 in / 12.45 mm
8. 0.56 in / 14.22 mm
9. 0.63 in / 16.00 mm
10. 0.70 in / 17.78 mm
11. 0.77 in / 19.56 mm
12. 0.84 in / 21.34 mm

Fig. 2–32 Tool Steel Ring

Instructions

Place the Tool Steel Ring on a non-ferrous central conductor (1.00 to 1.25 in / 2.5 to 3.2 cm diameter). Clamp the central conductor between head and tail stocks. Set the equipment amperage to the pre-determined levels according to AS 5282. Magnetize and apply particles according to normal procedures. Verify the number of indications observed against the minimum number required by AS 5282. If multiple amperages are being checked, proceed from the lowest to the highest setting, demagnetizing the ring between each setting.

It is recommended to benchmark the performance of the Tool Steel Ring on a specific magnetic particle equipment in order to detect any decrease or change in performance over time.

For best results, demagnetize ring after use, clean with acetone or equivalent cleaning solvent. Apply oil or other rust protection before storage.

Note: The Tool Steel Ring is not intended for use with AC magnetic fields.

■ 〖Point 5〗 Magnetic Field Indicator (Fig. 2–33)

Fig. 2–33 Indicator

General description

The Magnaflux Magnetic Field Indicators, also known as gauss meters or magnetometers, are used to check residual magnetism after magnetic particle testing. They read the amount of residual magnetism left in a part quickly when the indicator arrow is placed against a magnetized part. The field indicators are available in both general purpose and non-calibrated and calibrated models and can be re-calibrated through Magnaflux Authorized Service Centers.

Note: Calibrated field indicators should be re-calibrated every 6 months.

Instructions

Place the field indicator near or directly against the object being tested. The lower rim of the indicator below the arrow is the most sensitive part of the meter and should be placed closest to the part being measured.

The indicator should be placed near a position on the part that exhibits flux leakage such as the end of a bar shaped part.

Magnetic polarity of the field is being measured by the direction of the pointer deflection on the center zero scale. A plus (+) indicates the meter has been presented to a North magnetic pole and a minus (–) to a South magnetic pole. The higher the reading, the stronger the magnetic field.

Readings in gauss relate only to the magnitude of external leakage fields and should not be misconstrued as the flux density within the part.

Note:
1. If you place the indicator in a field strong enough, it may throw it considerably off scale.
2. If your field indicator comes in contact with the field of a demagnetizing coil or within the effective field of a conductor carrying a heavy alternating current, it may become demagnetize.

General description

The Magnaflux Magnetic Pie Field Indicator (Fig. 2–34), also known as a pie gauge, is a device used as an aid in determining the direction of magnetic field for detection of discontinuities in ferrous materials. It is an octagon shaped piece made with a low retentive steel surface having eight bonded slices, similar to pieces of pie. The octagon shaped piece is mounted on a handle so the inspector can place it on the part in the area being magnetized.

Fig. 2–34 Pie Field Indicator

With an adequate amount of magnetizing current and proper particle application, the pie gauge will show indications in the same direction natural indications would appear.

Instructions

Place the pie gauge as flat as possible on the part with the pie section side down. Make sure it is in the area to be magnetized.

Apply the magnetic current while applying the magnetic particles using the continuous method.

If using dry particles, lightly blow off the excess while the magnetic current is still on.

Artificial indications will appear on the pie gauge in the same direction as the natural defects.

■ 〖Point 6〗 Centrifuge Tubes and Stand to Measure Settling Volume

Centrifuge Tubes (Fig. 2-35) are used to monitor the concentration of magnetic particles and the level of contamination in fluorescent and visible baths.

Bath Strength

The amount of magnetic particles per gallon of fluid in the Testing bath is called its strength or concentration. If the bath concentration is below recommended strength, weak particle indications will be produced or possibly no indication will appear; therefore, defects will not be detected. If there are too many particles in the bath, indications may be masked by heavy background buildup.

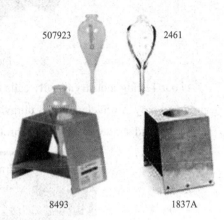

Fig. 2-35 Centrifuge Tubes

The usable limits of bath strength are quite broad, but for consistent results the bath strength should be maintained constant at all times.

A light bath strength usually forms good indications on deep cracks, but a heavier particle concentration will show fine defects better. The bath concentration, which will best detect all defects, should be determined and held constant. Bath strength should be checked at least once each day.

After the entire bath has been thoroughly mixed and agitated, it is essential to check it for strength. The most widely used method is by gravity settling in a graduated ASTM pear shaped centrifuge tube.

Daily instructions (including new bath):

(1) Let pump motor run for up to 30 minutes to agitate the suspension.

(2) Flow the bath mixture through hose and nozzle for a few moments to clear hose.

(3) Fill the centrifuge tube to the 100 mL line.

(4) Place the tube in the stand.

If required by written procedure, demage the tube (note that the stand is non-ferrous and will not interfere with particle settling) . Let the tube stand in a vibration-free area to allow particles to settle. Settling time is 30 minutes for a water bath or 60 minutes for an oil bath.

The gravity settling method applies to either oil or water suspension. In hot weather the water bath should be checked more often as is it more volatile than oil. Therefore, as water is lost by evaporation, it must be replaced. The settled particles (measured in mL) in the bottom of the tube indicate the amount of magnetic particles in suspension. A UV lamp, such as the Magnaflux EV6000 LED UV-A Lamp, must be used for fluorescent particles. Do not include dirt particles in your centrifuge tube readings. They usually settle on the top of the magnetic particles. Dirt will not fluoresce under UV-A irradiation. In visible particles, the appearance of dirt is very different than that of the particles. Dirt will be coarser and irregular in shape. See illustrations for recommended settling volume.

〖Point 7〗 Black Light (Fig. 2-36)

Fig. 2-36   Black Light

As our Testing industry usually calls black light, UV-A lamp, also known as ultraviolet lamp. Yet it produces only A wavelengths of ultraviolet light, filtering out harmful ultraviolet UV-B and UV-C, from visible and shorter wavelengths through filters (Fig. 2-37).

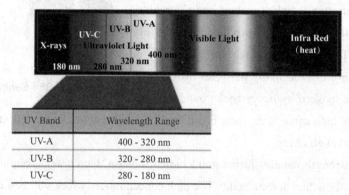

UV Band	Wavelength Range
UV-A	400 - 320 nm
UV-B	320 - 280 nm
UV-C	280 - 180 nm

Fig. 2-37   UV Band

Fluorescent penetrators and phosphors contain fluorescent dyes, a fluorescent dye absorbs energy from UV-A radiation, can stimulate the production of human eye recognizable yellowish green light (wavelength 550 nm). Particularly the UV-A, with a wavelength of 365 nm can cause most fluorescent dyes to excite the strongest fluorescence (yellowish green light). The human eye is most sensitive to yellowish green. Experimental results show that: the contrast rate between black dye and white imaging agent background is 9 to 1; The contrast rate between red dye display and white imaging agent is only 6 to 1; Fluorescence can be compared to a dark background of 300 to 1, Even 1 000 to 1. Therefore, Fluorescent osmotic detection and phosphor detection using black light have high detection sensitivity.

The precautions for the use of the black light are:

(1) The black light has just been lit, the output can not reach the maximum value, so the testing work should wait 5 min later.

(2) To minimize the number of lights on and off, frequent start will shorten the life of the black light.

(3) The black light is used, the radiation energy decreases, so the black light irradiance should be measured regularly.

(4) Voltage fluctuation of power supply has great influence on black light. The voltage is low, the lamp may not start or the lit lamp will go out; when the voltage used exceeds the rated voltage of

the lamp, it will have a great impact on the service life of the lamp, so the power supply should be installed when necessary to keep the supply voltage stable.

Put into Practice

1. Please look at Table 2-3, indicating the existing magnetic powder testing equipment type, model and manufacturer.

Table 2-3　The Company's Equipment

Equipment name	Equipment type	Technique parameter	Manufacturer
Magnetic flaws detector	CJX-220E	Yoke AC	Wuxi Jiecheng
Magnetic paste	BM-1	/	Wujiang Nan Ma
Contrast intensifying agent	FA-5	/	MARKTEC
Magnetic suspension liquid	MT-BO	/	MARKTEC
Cleaner	DPT-5	/	MARKTEC
Penetrant	DPT-5	/	MARKTEC
Developer	DPT-5	/	MARKTEC
Light meter	TES-1330A	/	Tanwan Tai shi

2. Look at Table 2-4, answer what the object is and how long the check cycle is.

Table 2-4　Equipment Check Table

Test	Interval
System performance	1 day
Magnetic ink concentration	8 hours of operation
Magnetic ink condition	1 week
Water break test	1 day
Ammeter calibration	1 year
Magnetic field indicator	3 months
Dead weight check	
— electromagnets	6 months
— permanent magnets	1 month
Visible/black light intensity	1 week

3. Read the following essay and answer how long the human eye (Fig. 2-38) has been adapted to dark environments, and what color is the most sensitive to the human eye in terms of visual functions?

Just as lighting is an important consideration in the inspection process, so is the eye's response to light. Scientists have recently discovered that a special, tiny group of cells at the back of the eye help tell the brain how much light there is, causing the pupil to get bigger or smaller. The change in pupil diameter is not instantaneous, therefore, eyes must be given time to adapt to changing lighting conditions. When performing a fluorescent magnetic particle inspection, the eye must be given time to adapt to the darkness of the inspection booth before beginning to look for indications. Dark adaptation time of at least one minute is required by most procedures. Some studies recommend adaptation time of five minutes if entering an inspection area from direct sunlight. Inspectors should carefully adhere to the required adaptation

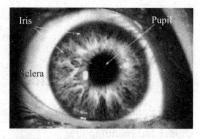

Fig. 2-38 Eye

time as it is quite easy to overlook an indication when an inspection is started before the eyes have adjusted to the darkened conditions.

Ocular fluorescence

When ultraviolet light enters the human eye, the fluid that fills the eye fluoresces. This condition is called ocular fluorescence and while it is considered harmless, it is annoying and interferes with vision while it exists. When working around ultraviolet lights, one should be careful not to look directly into lights and to hold spot lights to avoid reflection. UV light will be reflected from surfaces just as white light will, so it is advisable to consider placement of lights to avoid this condition. Special filtered glasses may be worn by the inspector to remove all UV light from reaching the eyes but allowing yellow-green light from fluorescent indications to pass. Technicians should never wear darkened or photochromatic glasses as these glasses also filter or block light from fluorescent indications.

2.4 Magnetic Particle Testing Procedure

〖Point 1〗 Magnetic Particle

For magnetic powder (Fig. 2-39), its function is to display medium. First, black magnetic powder, its composition is iron trioxide (Fe_3O_4), black powder, suitable for light or bright background of the workpiece.

Followed by red magnetic powder, whose composition is iron trioxide (Fe_2O_3), which is iron-red powder-like and suitable for dark background workpieces (Fig. 2-40).

Fig. 2-39 Magnetic Powder

Then there is the phosphor, which is wrapped in the particles of ferromagnetic powder trioxide and can emit yellowish green fluorescence under ultraviolet radiation. It is suitable for dark background workpieces, especially for the reason of human eye sensitivity. The magnetic powder test with phosphor as magnetic medium has higher sensitivity than other magnetic powder.

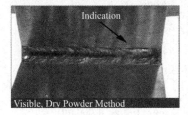

Fig. 2-40　Indication

〖Point 2〗 Magnetic Ink

The commonly used magnetic suspensions are water magnetic suspensions and oil magnetic suspensions. The viscosity values of the two magnetic suspensions are different and their fluidity is not consistent, which leads to the difference of detection sensitivity.

1. Preparation of non-fluorescent oil magnetic suspensions (including black, red, white, blue)

(1) Non-fluorescent oil magnetic suspension dispersant, kerosene or tasteless kerosene mixed with transformer oil or $10^{\#}$ oil.

The dispersant ratio is shown in Table 2-5.

Table 2-5　Non-fluorescent Qil Magnetic Suspension

Formula number	Material name	Proportion (%)
1	odorless kerosene	100
2	kerosene+electric insulating oil	50 +50
3	electric insulating oil	100
4	kerosene+$10^{\#}$ engine oil	50 +50

(2) Magnetic powder solution should be suitable, generally adding 15-25 g of magnetic powder per liter of dispersant.

(3) A small amount of oil is mixed with magnetic powder to make the magnetic powder wet, stir into a uniform paste state and then add other oil.

2. Preparation of non-fluorescent magnetic suspensions (including black, red, white, blue)

(1) Non-fluorescent hydro magnetic suspension formulation(Table 2-6).

Table 2-6　Non-fluorescent

Formula number	Elements	Weight or proportion	Magnetic powder content
1	YF-3 dispersant, sodium nitrite, water	in proper order: 2%, 1%, 1 000 mL	15-25 g

continued

Formula number	Elements	Weight or proportion	Magnetic powder content
2	Soap, sodium nitrite, water 50 ℃ -60 ℃	in proper order: 5 g, 5-15 g, 1 000 mL	15-25 g
3	Magnetic paste, water	in proper order: 60-80 g, 1 000 mL	-
4	100$^\#$ concentrated milk, triethanolamine, sodium nitrite, defoamer, water	in proper order: 10 g, 5 g, 5 g, 1-2 g, 1 000 mL	15-25 g

(2) Preparation method:

$1^\#$ formula—Mixing the magnetic powder dispersant YF-3 evenly and then weighing it out according to the dosage. First dilute with a small amount of water and add magnetic powder to stir until completely wet, then add a small amount of water to dilute, add sodium nitrate, stir well and add the rest of the water to mix well before use.

$2^\#$ formula—Take a small amount of water to dissolve soap, then add appropriate amount of water and sodium nitrate and magnetic powder to stir well, then add the rest of the water to mix well.

Formula—Add 100$^\#$ concentrated milk to 1 liter 50 degrees of warm water, stir until completely dissolved, then add triethanolamine, sodium nitrite and defoamer, and stir well after adding one ingredient.

When adding magnetic powder, a small amount of dispersant is mixed with magnetic powder to make the magnetic powder wet and then other dispersants are added.

3. Preparation of fluorescent oil magnetic suspension

The phosphor is made of magnetic iron oxide powder, industrial pure iron powder, hydroxyl iron powder and so on.

Compared with non-fluorescent magnetic powder, fluorescent magnetic powder is yellowish green fluorescence excited by ultraviolet light, bright and easy to observe, good visibility and contrast, and clearer display of parts defects. Used on any color surface (with ultraviolet light source).

The use of fluorescent magnetic powder can improve the testing speed, effectively reduce the leakage rate and reduce the eye fatigue of the staff.

When preparing fluorescent magnetic powder oil suspension, odorless kerosene should be used as dispersant, but not other kerosene which can emit fluorescence itself, because on the one hand, this kerosene will produce fluorescence on the surface of the workpiece and interfere with the display of defects. On the other hand, it will reduce the luminous intensity of phosphor.

The preparation ratio is 1-2 g phosphor per liter of tasteless kerosene.

Preparation method: wetting magnetic powder with a small amount of tasteless kerosene, then adding the rest of kerosene.

4. Preparation of fluorescent hydro-magnetic suspensions

The moisture dispersible agent for the preparation of magnetic suspension of fluorescent magnetic powder should be strictly selected. In addition to meeting the performance requirements of water dispersion agent, the phosphor powder should not be agglomerated, dissolved and deteriorated.

Suggested YC-2 phosphor may use YF magnetic powder or the following formula:

Emulsifier (JFC): 5 g, sodium nitrite: 15 g, defoamer ($28^{\#}$): 0.5–1 g, phosphor: 1–2 g, water: 1 liter.

Preparation method: The emulsifier and defoamer are stirred evenly, and the sufficient water is added in proportion to become a water dispersible agent. A small amount of water dispersant and magnetic powder are stirred evenly, then the residual dispersant is added and then sodium nitrite is added.

The concentration of magnetic suspension refers to the number of grams of magnetic powder per liter of liquid.

If the concentration is too low, the small defect will be missed; if the concentration is too high, the lining will be reduced, and excessive magnetic powder will be attached to the magnetic pole of the workpiece, which will interfere with the display of the defect, so the preparation concentration should be appropriate.

In the process of magnetic particle testing, each workpiece should adsorb a certain amount of magnetic powder after magnetization. Therefore, after using magnetic suspension for a period of time, the concentration of magnetic suspension should be determined. In order to ensure the detection accuracy and reliability of magnetic powder testing.

The most accurate way for users to detect the concentration of magnetic suspensions is to use magnetic powder precipitates.

For very experienced operators or users who do not buy magnetic powder precipitators, they can also be judged by visual methods based on experience.

(1) The color of magnetic suspension changes greatly, there are a lot of dirt, should be remodulated magnetic suspension.

(2) The user in flaw detection, the surface of the workpiece magnetic powder distribution is light or there is no magnetic mark accumulation. Consideration should be given to adding magnetic powder or remodulating magnetic suspensions.

■ 〖Point 3〗 Operation Flow and Evaluation of Magnetic Particle Testing

1. Pretreatment

All materials and specimens should be free of grease and other impurities that may affect the normal distribution of magnetic powder, the density, characteristics and clarity of magnetic powder deposits.

2. Magnetization

Magnetic particle testing shall be based on ensuring satisfactory detection of any harmful defects. Make the magnetic line cross any defects that may exist in the specimen within a feasible range.

3. Apply magnetic powder or magnetic suspension

Magnetic suspension should be fully stirred before each application. When the magnetic powder suspension is poured on the surface of the sample, the defective magnetic marks on the surface of the specimen can be seen in sunlight and these magnetic marks will be clearer under ultraviolet radiation, especially in the case of black light. The method of manual spraying and continuous detection is adopted, that is, the magnetization of the tested parts, the application of magnetic powder and the observation of magnetic marks should be completed within the magnetization electrification time, the electrification time is 1–3 s, and the magnetic suspension is stopped at least 1 s before the magnetization can be stopped.

4. Magnetic mark observation and recording

Defect magnetic mark observation should be carried out immediately after the formation of magnetic mark. Except to confirm that the magnetic mark is caused by the local magnetic unevenness or improper operation of the workpiece material, the other magnetic mark display should be treated as a defect. Apply 2–10 times magnifying glass to identify small magnetic marks. The magnetic mark display record is recorded by photography and marked on the photo.

5. Demagnetization treatment

The workpiece can be demagnetized locally by AC yoke or by winding coil. The demagnetization effect of the workpiece is generally measured by magnetic field intensity meter, not more than 0.3 mT (240 A/m).

6. Evaluation of results (Table 2-7)

(1) No cracks and white spots are allowed.

(2) Quality grading.

(3) Comprehensive rating.

Table 2-7　Magnetic Testing Acceptance Criteria for Structural Steel Welds

Type of defect	Structural category		
	Special	Primary	Secondary
Cracks	Not acceptable		
Incomplete penetration or lack of fusion	Not acceptable		On the root side of welds for which back welding is not required; length $< t/2$, maximum 10 mm and not closer than t
Surface porosity	Not acceptable		Not acceptable in areas with tensile stresses, In other areas, the accumulated pore diameters in any area of 10 mm×15 mm are not to exceed 15 mm, maximum size of single pore; $t/4$ or 4 mm (whichever is the smaller)
Undercut, maximum depth, mm	Not acceptable		Not acceptable when transverse to tensile stresses, maximum depth allowed in other areas 0.75 mm
t is the nominal plate thickness. General requirements: Welds shall be of correct shape, size and geometry. Welds shall have a regular finish and merge smoothly into the base material, groove welds shall have slight or minimum reinforcement or root penetration not exceeding 3 mm in height. The face of fillet welds shall be slightly convex or concave or flat and leg lengths shall be equal.			

When there are many defects in the circular defect evaluation area, a comprehensive rating should be carried out. For all kinds of defects, the lowest grade of quality grade is taken as the grade of comprehensive rating, and when the grade of each kind of defect is the same, the first grade is reduced as the grade of comprehensive rating. The detailed testing procedure is shown in Fig. 2–41.

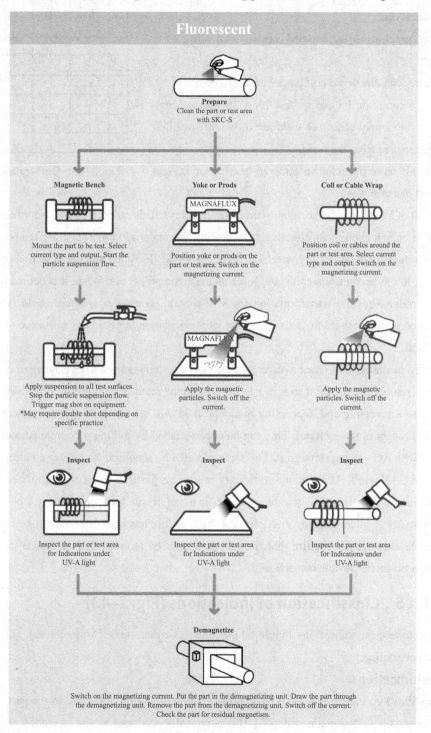

Fig. 2–41　Testing Procedure

〖Point 4〗 Demagnetization (Fig. 2–42)

After conducting a magnetic particle inspection, it is usually necessary to demagnetize the component. Remanent magnetic fields can:

•Affect machining by causing cuttings to cling to a component.

•Interfere with electronic equipment such as a compass.

•Create a condition known as "arc blow" in the welding process. Arc blow may cause the weld arc to wonder or filler metal to be repelled from the weld.

•Cause abrasive particles to cling to bearing or faying surfaces and increase wear.

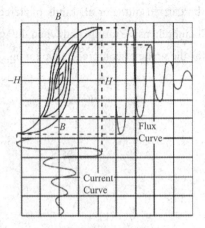

Fig. 2–42   Demagnetizition

Removal of a field may be accomplished in several ways. This random orientation of the magnetic domains can be achieved most effectively by heating the material above its curie temperature. The curie temperature for a low carbon steel is 770 ℃ or 1 390 °F. When steel is heated above its curie temperature, it will become austenitic and loses its magnetic properties. When it is cooled back down, it will go through a reverse transformation and will contain no residual magnetic field. The material should also be placed with it long axis in an east-west orientation to avoid any influence of the earth's magnetic field.

It is often inconvenient to heat a material above its curie temperature to demagnetize it, so another method that returns the material to a nearly unmagnetized state is commonly used. Subjecting the component to a reversing and decreasing magnetic field will return the dipoles to a nearly random orientation throughout the material. This can be accomplished by pulling a component out and away from a coil with AC passing through it. The same can also be accomplished using an electromagnetic yoke with AC selected. Also, many stationary magnetic particle inspection units come with a demagnetization feature that slowly reduces the AC in a coil in which the component is placed.

A field meter is often used to verify that the residual flux has been removed from a component. Industry standards usually require that the magnetic flux be reduced to less than 3 gausses after completing a magnetic particle inspection.

〖Point 5〗 Classification of Indications (Fig. 2–43)

Once detected, the indications should be classified as either "false" "nonrelevant", or "relevant" before final evaluation.

False Indications

False indications can be produced by improper handling and do not relate to the part's condition or use. An example is "magnetic writing" . This is typically produced by the formation of indications at local poles that are created when the part comes in contact with another magnetized part prior to or during testing. This can be eliminated by demagnetization and repeating the testing.

Magnetic writing is most likely to occur when using the residual method, through poor handling that allows the individual parts to touch. The continuous technique may require the demagnetization of parts before the next testing to preclude the possibility of magnetized components touching. This type of false indication can be eliminated through careful handling.

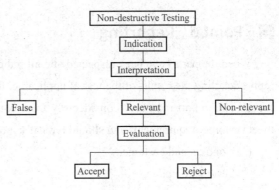

Fig. 2–43   Classification of Indications

Other sources of false indications may be caused through the use of excessively high magnetizing currents or inadequate precleaning of the parts to remove oil, grease, corrosion products and other surface contaminants.

Nonrelevant Indications

These are the result of flux leakage due to geometrical or permeability changes in the part. Examples of geometric causes include splines, thread roots, gear teeth, keyways or abruptsection changes . A concern with these conditions is that they may also be stress risers and could be the origin for fatigue-induced cracks. These conditions are therefore some of the most critical; the possibility that one of these nonrelevant indications can conceal a crack must be considered. Other potential sources of nonrelevant indications include localized permeability changes in the part, which may be due to localized heat treatment or variations in hardness, and may also occur at the fusion zone of a weld.

Relevant Indications

These are produced by flux leakages due to discontinuities in the part. When these discontinuities are not in compliance with a code, they are classified as rejectable. If they meet the acceptance criteria they are considered to be acceptable discontinuities. Discontinuities that do not permit the part to be used for its original purpose or can potentially cause the part or fail are classified as defects.

Indications produced during are shown in Fig. 2–44.

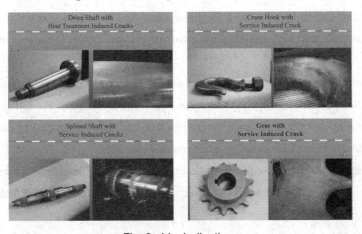

Fig. 2–44   Indications

〖Point 6〗 Reporting

When the parts have been inspected and all indications evaluated, it will be necessary to prepare a report detailing the results of the test, if applicable, the size, location and orientation of discontinuities found. This report may vary considerably from company to company, but as a minimum, it should meet customer requirements and should typically include the following data:

(1) Contract and customer.

(2) Testing company.

(3) Date of testing.

(4) Inspector's name and qualification and certification level.

(5) Report number or identification.

(6) Applicable codes, specifications or procedures, including type and technique of in Section.

(7) Acceptance criteria.

(8) Component description, part number and serial number.

(9) Flux line direction with respect to discontinuity orientation.

(10) Other identification details as requested in the contract; for example, batch or order number.

(11) Material batch number (particles, liquid carrier, etc.).

(12) Results of testing, including recording of indications as detailed below.

(13) Signature or testing stamp of inspector.

A detailed, concise report will enable future evaluations by other inspectors. There are several ways of achieving this:

(1) A descriptive written report including significant dimensions and indication location can be complied.

(2) A photograph may be taken under the correct viewing conditions. Black light illumination in a darkened environment works fine if the correct exposure (usually several seconds) is used and a camera is mounted on a tripod. It may be necessary to vary the exposure time several times to ensure that the best lighting and exposure parameters are met.

(3) A free-hand sketch may be used to supplement the data. Again, key dimensions must be included. Unfortunately, not everyone is a good artist and the quality and usefulness of sketches can be questionable.

(4) A piece of transparent tape may be used to lift the indication from the test surface if using dry particles. When peeled off, the tape will retain the shape and size of the indication through the adherence of the particles to its adhesive layer. The tape can then be applied to the report or other suitable background material to render the indication more visible.

(5) Aerosol-based, strippable lacquer may be applied in several thin layers, allowing each layer to dry before applying the next, until sufficient thickness exists, allowing the solidified, flexible film to be peeled off the part.

(6) Magnetic rubber testing may be used to create a permanent record of indications. Magnetic

rubber testing involves using a two-part liquid rubber kit consisting of :

① Room temperature vulcanizing (RVT) rubber supplied in a liquid form. This liquid rubber also contains ferromagnetic powder.

② A catalyst, which when mixed will cause the mixture to solidify at a controlled rate. When mixed together, the rubber solution is poured onto the area of testing and a magnetizing force applied during the "cure" time. The cure rate should be slow enough to allow the ferromagnetic powder to migrate to the flux leakage fields, but not so slow as to delay the testing longer than necessary. Magnetic rubber materials with curing ranges from 5 minutes to 4 hours are available. When the rubber is solidified, it can be peeled off and the indications created by the flux leakage can be observed and retained as a record.

The magnetic rubber testing technique has uses beyond recording indications. It can also be used to inspect areas and surfaces that are inaccessible for standard MT Testing, such as inside blind fastener holes, particularly threaded holes. The rubber can be poured into the hole and a permanent magnet applied across the hole during the cure. After solidification of the rubber, the solid (but flexible) rubber plug is removed.

From the hole and the indications from within the hole can be viewed on the outer surface of the plug. A small stick or other device can be cast into the plug to facilitate easy removal.

The magnetic particle test method is effective for the detection of surface and slightly subsurface discontinuities in ferromagnetic parts. It can be used as a testing tool at all stages in the manufacture and end use of a product.

Put into Practice

1. Read Table 2-8 and answer the questions that show how to classify by size and what the criteria are for each level.

Table 2-8 The Criteria Are for Each Level

Type of indication	Acceptance level		
	1	2	3
Linear indication l=length of indication	$l \leqslant 1.5$ mm	$l \leqslant 3$ mm	$l \leqslant 6$ mm
Non-linear indication d=major axis dimension	$d \leqslant 2$ mm	$d \leqslant 3$ mm	$d \leqslant 4$ mm

Acceptance level 2 and 3 may be specified with a suffix "X" which denotes that all linear indications detected shall be assessed to level t. However, the probability of detection of indications smaller than those denoted by the original acceptance level can be low.

2. Read the following essay (Fig. 2-45) to answer where magnetic powder detection should focus on safety.

SAFETY

Safety must be observed at all times and should not be restricted to the followings:

(1) Observe all warning signs, notices and regulations at work site.
(2) Dressed in proper working attires such as coverall, safety helmet, safety boots, etc.
(3) When working in confine area, ensure ventilation fan with adequate supply of fresh air.
(4) Ensure confine area is gas free.
(5) Ensure fire extinguisher is available.
(6) Ensure magnetizing unit is free from leakage of electricity before attempting examination.
(7) Wear protective gloves, clothing etc., if examination medium/chemical is sensitive to skin or eyes.
(8) Do no accidentally aim jet of Aerosol Can to your eyes or other personnel.
(9) Remove and discard in proper place all dirty rags and other used items and exhausted Aerosol Can.

Fig. 2-45 Safety

3. Translate the following inspection reports (Table 2-9).

Table 2-9 PT Report

Attachment 2-PT report							Page /
Period of examination Date	Technician 1 (Init.)	Cert no. tech 1	Assistant 2 (Init.)	Cert no. Ass 2	Assistant 3 (Init.)	Ceit no. Ass 3	
Assistant 4 (Init.)	Ceit no. Ass 4	Assistant 5 (Init.)	Ceit no. Ass 5	Examinalion organization/Inspection authority			
Client		Contractor		Location			
Object				Section ID-no.	Drawing no.	ECN no.	
Examination procedure		Acceptance criteria		System manufactuerr			
Supplementary information (without Hability)							
Weld preparation				Other	Heart-treated	Tine of examination	
☐X ☐K	☐Y ☐I	☐U ☐Fillet	☐V	☐	☐Yes ☐No	Hours after weld	
Welding process	121	131	135	136	141	Other	
☐Mane. lec. 111	☐Sub.arc	☐Mig	☐Mag	☐Fluxcored wire	☐TIG	☐	
Surface condition					Reinforcement		
☐Fine	☐Smooth	☐General	☐Primer coated		☐As weled	☐Ground	
Examination procedure					Other		
☐EN 571-1	☐ASME V, art.6	☐ISO 3452			☐		
Precleaning				Other	Cleaning media type.		
☐None ☐Liquid solvent ☐Rinsing bath ☐Mechanical				☐			
Pemetrant system							
Name			Manufacturer				

continued

Penetrant type				Batch no.
☐ I. fluorescent	☐ II. color contrast			
Cleaning of excess penetrant				Batch no.
☐ A. Water washable ☐ B. lipophilic ☐ C. Solvents ☐ D. Hydrophilic emulsifier ☐ E. Water and solvent				
Developer				Batch no.
☐ A. Dry ☐ B. Water soluble ☐ C. Water suspendable ☐ D. Solvent based				
Penetration time	Emulsifier contact time	Develop time		Test piece temp.
Min.	Min.	Min.		℃
Lighting				Serial no. lamp
☐ White light Lux ☐ UV–Light μW/m²				
Acceptance criteria		Incl. virual exam	Other	
☐ EN 1289, level		☐ EN ISO 5817, Weld class	☐	
Extent of examination				
Results of examination				
Repairs		Enalosures	Technician (Date/Signature)	
☐ Marked on object ☐ Marked on sketch ☐ Checked after grinding		Sheets		

Words and Phrases

magnetite ['mægnɪˌtaɪt] n. 磁铁矿
cannon ['kænən] n. 大炮,加农炮
boiler ['bɔɪlə] n. 汽锅,锅炉
in-service 在役的
overhaul [ˌəʊvə'hɔːl] v. 检查
naked eye (unaided eye) 肉眼
calibrate ['kælɪˌbreɪt] v. 校准

demagnetization [diːˌmægnətaɪ'zeɪʃən] n. 除去磁性,退磁
remnant ['remnənt] n. 残余,剩余 adj. 剩余的,残留的
magnetism ['mægnɪˌtɪzəm] n. 磁,磁学
ferromagnetic [ˌferəʊmæg'netɪk] adj. 铁磁的,铁磁体

paramagnetic [ˌpærəmæg'netɪk] adj. 磁性的
diamagnetic [ˌdaɪəmæg'netɪk] adj.［物］反磁性的，逆磁性的反磁性体
nickel ['nɪkl] n.［化］镍（化学元素）
austenitic stainless steel 奥氏体不锈钢
bismuth ['bɪzməθ] n.［化］铋
antimony ['æntɪmənɪ] n. 锑（化学元素）
polarity [pəʊ'lærətɪ] n. 极性
repulsion [rɪ'pʌlʃn] n. 推斥，排斥
magnetic field n. 磁场
lines of force 磁力线
flux density 磁通密度
magnetograph [mæg'niːtəʊgrɑːf] n. 磁强记录仪
magnetic flux 磁通量
closed loop 闭环，密闭回路
flux density 通量密度
gauss [gaʊs] n. 高斯（磁感应强度单位）
tesla ['teslə] n. 特斯拉（磁通密度单位）
permeability [ˌpɜːmɪə'bɪlətɪ] n. 磁导率，渗透性
magnetic permeability 磁导率
weber ['veɪbər] n. 韦伯（磁通量单位）
magnetic domain 磁畴
root mean square (RMS) 均方根
skin effect 集肤效应
permanent magnet 永久磁铁
misnomer [ˌmɪs'nəʊmə] n. 误称，用词不当
tubular ['tjuːbjʊlə] adj. 管状的
distillate ['dɪstɪleɪt] n. 精华，蒸馏物
flammability [ˌflæmə'bɪlətɪ] n. 易燃，可燃性
headstock ['hedˌstɒk] n. 主轴箱
haphazard [hæp'hæzəd] adj. 偶然的，任意的
relative permeability 相对磁导率
International Annealed Copper Standard (IACS) 国际退火铜标准
nonmagnetic [ˌnɒnmæg'netɪk] adj. 无磁性的
magnetic reluctance 磁阻
saturate ['sætʃəreɪt] v. 使饱和

magnetic saturation 磁（性）饱和
sustained [sə'steɪnd] adj. 持续不变的，相同的
provided that 假如
leakage field 漏磁场
liquid suspension 悬浮液
magnetic ink 磁悬液
hysteresis loop 磁滞回线
induced magnetic flux density (B) 磁感应强度
magnetizing force (H) 磁场强度，磁化力
dashed line 虚线
remanence ['remənəns] n. 顽磁，剩磁
electromagnetic yoke 电磁轭
curie point 居里点
longitudinal magnetization 纵向磁化
threader bar 穿棒（法）
amperage ['æmpərɪdʒ] n. 安培数，电流强度
retentivity [ˌriːten'tɪvɪtɪ] n. 保持力
coercivity [ˌkəʊɜː'sɪvətɪ] n. 矫顽（磁）力，矫顽（磁）性
coercive force n. 抗磁力，矫顽（磁）力
abrasive [ə'breɪsɪv] n. 研磨剂 adj. 研磨的
circular magnetism 周向磁化（磁力线成圆形）
rectification [ˌrektɪfɪ'keɪʃn] n. 整流，矫频，检波，调整
half-wave rectification 半波整流
full-wave rectification 全波整流
tailstock ['teɪlˌstɒk] n.［机］尾座，尾架，顶针座
pneumatic [nju'mætɪk] adj. 装满空气的，有气胎的，气力的，风力的气胎
contour ['kɒntʊə] n. 轮廓，周线，等高线
ballast transformer 镇流变压器
ultraviolet ['ʌltrə'vaɪələt] adj. 紫外线的，紫外的，紫外线辐射
black light 黑光
gauge [geɪdʒ] n.（量，卡，线）规，计，（仪）表
quantitative quality indicator (QQI) 定量磁场指示器

kerosene ['kerəsi:n] *n.* 煤油
petroleum [pə'trəʊliəm] *n.* 汽油
octagonal [ɒk'tægənl] *adj.* 八边形的，八角形的
groove [gru:v] *n.* 凹槽

A Sheet Work Manual

Magnetic Particle Inspection Procedure

1. Scope

(1) This procedure give the general requirements for carrying out the examination of welds in ferric materials using the AC electro magnetic yoke technique.

(2) This procedure meets the requirements of ASME Section V Article 7 and ASTM E-709-95.

2. Description of method

(1) This method involves the magnetisation of an area to be examined and the application of ferromagnetic particles to the surface. The particles gather at areas of magnetic flux leakage and form indications characteristic of the type of discontinuity detected.

(2) Maximum sensitivity is achieved when linear discontinuities are oriented perpendicular to the lines of flux.

(3) The AC electromagnetic yoke technique is restricted to the detection of surface breaking discontinuities.

3. Equipment and materials

This procedure is intended for use with the following equipment and consumables or their equivalent.

- Magnaflux Y6 AC Yoke
- Portable ultra violet light
- Magnaflux WCP-1 White contrast paint
- Magnaflux WCP-2 White contrast paint
- Magnaflux 7HE Black Ink
- Magnaflux SKC-NF cleaner
- Magnaflux SKC-S cleaner
- Magnaflux dry powder (red, yellow or grey colour)
- Magnaflux 14 HF Fluorescent Ink
- Magnaflux 20 A Fluorescent Ink concentrate (water based)
- Burmah Castrol magnetic field indicators Type 1 Brass finish.

4. Parts to be examined

Welds in ferritic materials, whether in the as welded or dressed condition and the associated heat affected zones and parent material within at least one inch of the weld on both sides of the weld.

5. Surface preparation

(1) Prior to the test the area to be inspected and at least one inch either side shall be free from

any features that may inhibit the test or mask unacceptable discontinuities. These include but are not limited to slag, spatter, oil, scale, rough surface and protective coatings.

(2) Surface preparation by grinding, machining or other methods may be necessary where surface irregularities could mask indications of unacceptable discontinuities.

(3) The temperature of the test surface shall not exceed 135 °F for magnetic inks and 600 °F for dry powders (For dry powders, test surface be clean and dry).

(4) For parts to be inspected using magnetic inks the area to be inspected may, if necessary, be precleaned with a cloth lightly moistened with cleaner.

(5) Where parts are to be examined using powders or fluorescent inks, the surface finish as detailed in 5 (1) to 5 (4) is adequate.

(6) When using black magnetic inks, the surface may be given contrast enhancement by applying a thin, even coating of white contrast paint of a type as detailed in Section 3 (1).

6. Equipment and consumable control

(1) The magnetizing force of yokes shall be checked at least once a year or after any damage and/or repair, the yoke shall be able to lift a weight of at least 10 pounds at the maximum pole spacing that will be used.

(2) Magnetic powders shall be used on a once only, expendable basis. Care shall be taken to avoid possible contamination.

(3) Magnetic inks are also once only, expendable materials and care shall be taken to thoroughly agitate the ink before use.

(4) Magnetic inks mixed from concentrates shall be subject to a settlement test before use. Settlement time shall be 30 minutes and settlement volume for the solids shall be as below:
- Fluorescent ink   0.1%–0.4%
- Non-fluorescent ink   1.2%–2.4%

(5) The black light intensity at the examination surface (15 in from the face of the light lens filter) shall be not be less than 1 000 $\mu W/cm^2$. Any increase in this value to suit client's specific requirements shall be detailed on the technique sheet. The bulb shall be allowed to warm up for at least 5 minutes before use.

(6) The black light intensity shall be checked at least every 8 hours of use in accordance with Secion 6 (5) using a calibrated black light meter.

7. Lighting conditions

(1) When conducting an examination under white lighting conditions, the inspector shall ensure that the level of lighting is adequate at the surface of the part (recommended minimum 100 foot candles or 1 000 lx). 50 foot candles (500 lx) lighting may be used if agreed with customer, for field inspections.

(2) Examinations under ultra violet lighting should have a background white light level of less than 20 lx and an ultra violet intensity at the test surface of not less than 1 000 $\mu W/cm^2$. Any increase in these values shall be specified in the technique sheet.

(3) The inspector shall allow at least 5 minutes for dark adaptation before beginning the inspection.

(4) If the examiner wears glasses or lenses, they shall not be photosensitive.

8. Direction of magnetizing field

(1) The magnetizing field shall be applied sequentially in two directions approximately perpendicular to one another.

(2) The direction of the field may be determined by using the Burmah Castrol magnetic field indicator. This will give its strongest indications when placed across the flux direction.

(3) Determination of field direction shall be carried out for each geometry of weld to be inspected.

9. Sequence of operations

(1) The surface to be inspected shall be prepared as Section 5.

(2) The continuous magnetisation technique is to be employed.

(3) Ensure in all operations that the pole faces remain in maximum contact with the surface.

(4) Position the poles as described in Section 8 and turn on the magnetizing field.

(5) For the wet magnetic particle method, apply the ink onto the area under test and allow to flow over the surface such as to allow maximum exposure of the magnetic particles to any flux leakage present, excess material may be gently blown across the surface to aid interpretation.

(6) Dry magnetic powders should be applied in such a manner that a light, uniform, dust like coating settles on the surface of the area under inspection. Excess power may be gently blown across the surface to aid interpretation and increase exposure of magnetic particles to any flux leakage present.

(7) Maintain the magnetizing field for at least two seconds after step 9 (5) or 9 (6) and inspect immediately.

(8) Repeat the above sequence at approximately 90% to the above.

(9) Repeat the above steps to cover the complete weld area under inspection ensuring an overlap between inspected areas of at least 25% of the pole spacing.

(10) The area to be inspected shall be limited to a maximum distance of 1/4 of the pole spacing on either side of line joining the two legs.

(11) Pole area to be inspected shall be limited to a maximum distance of one fourth of the pole spacing on either side of line joining the two legs.

(12) Pole spacing shall be limited to a maximum value equal to or less than that used when conducing the lift test of the standard weight but shall be not less than 3 inches.

10. Evaluation of indications

(1) An indication may be the evidence of a surface imperfection. All indications need not be relevant. Relevant indications are those caused by leakage flux. Relevant indications due to unacceptable mechanical discontinuities are to be noted, located and sized.

(2) Any indication which is believed to be non relevant shall be re-examined again. Only indications having major dimension of greater than 1/16 of an inch shall be considered relevant.

(3) A liner indication is one having a length greater than its width. A rounded indication is shape with a length equal to or less than 3 times its width.

(4) Unacceptable indications shall be removed by chipping or grinding and shall be re-tested, if the repaired surface is free of sharp notches are corners/grinding marks. When a defect appears to be fully removed the area can be repaired and re-examined by the same method. Repaired area shall be blended into the surrounding surface so as to avoid sharp notches, crevices or corners.

(5) If required the repaired area shall also be re-examined by another suitable NDT method.

11. Acceptance criteria

The acceptance criteria given hereunder for relevant indications for various codes shall be used in general. The inspector using this is cautioned that Code requirements do change and that in the case of conflict it shall be referred (in writing) to the appointed NDT Level Ⅲ or to the Divisional Manager (NDT).

(1) ANSI/ASME B31.1, Power Piping.

① Evaluation of Indications.

Mechanical discontinuities at the surface will be indicated by the retention of the examination medium. All indications are not necessarily defects since certain metallurgical discontinuities and magnetic permeability variations may also produce similar indications which are not relevant to the detection of unacceptable discontinuities.

Any indications which are believed to be non-relevant shall be re-examined to verify whether or not actual defects are present. Surface conditioning may precede the re-examination. Non-relevant indications which would mask indications of defects are unacceptable.

Relevant indications are those which result from unacceptable mechanical discontinuities. Linear indications are those indications in which the length is more than three times the width. Rounded indications are indications which are circular or elliptical with the length less than three times the width.

An indication of a discontinuity may be larger than the discontinuity that causes it; however, the size of the indication and not the size of the discontinuity is the basis of acceptance or rejection.

② Acceptance Standards.

The following relevant indications are unacceptable:

Any cracks or linear indications.

Rounded indications with dimensions grater than 3/16 in. (5.0 mm).

Four or more rounded indications in a line separated by 1/16 in. (2.0 mm) or less, edge to edge.

Ten or more rounded indications in any 6 sq. in of surface with the major dimension of this area not to exceed 6 in (150 mm) with the area taken in the most unfavourable location relative to the indications being evaluated.

(2) ASME Boiler and Pressure Vessel Code, Sec. VIII Div. 1 Pressure Vessels.

① Evaluation of Indications.

Indications will be revealed by retention of magnetic particles. All such indications are not necessarily imperfections, however, since excessive surface roughness, magnetic permeability variations (such as at the edge of Heat Affected Zones), etc. may produce similar indications. An indication is the evidence of a mechanical imperfection. Only indications which have any dimension greater than 1/16 in. shall be considered relevant.

② Acceptance Standards.

All surfaces to be examined shall be free of:

Relevant linear indications.

Relevant rounded indications greater than 3/16 in.

Four or more relevant rounded indications in a line separated by 1/16 in. or less, edge to edge.

(3) AWS D1.1　Structural Welding Code-Steel.

Inspections may be performed immediately after the completed welds have cooled to ambient temperature. Magnetic Particle Testing on welds in ASTM steels A514 and A517 shall be performed no sooner than 48 hours after completion of the weld.

Indications revealed by Magnetic Particle Testing shall be evaluated as per applicable categories as follows:

Statically Loaded Structures:

Acceptance criteria shall be as per 8.15.5 (i.e. 8.15.1) of AWS D1.1 (1992 Edition).

Dynamically Loaded Structures:

Acceptance criteria shall be as per 9.25.2 of AWS D1.1 (1992 Edition) .

Tubular Structures:

Acceptance criteria shall be as per 10.17.5 (i.e. 10.17.1) of AWS D1.1.

(4) API Standard 1104 Pipelines and Related Facilities.

① Classification of Indication.

Any indication with a maximum dimension of 1/16 in. (1.59 mm) or less shall be classified as non relevant. Any larger indication believed to be nonrelevant shall be regarded as relevant until re-examined by magnetic particle or another non-destructive testing method to determine whether or not an actual discontinuity exists. The surface may be ground or other wise conditioned before re-examination. After an indication is determined to be non-relevant, other non-relevant indication of the same type need not be re-examined.

Relevant indications are those caused by discontinuities. Linear indications are those in which the length is more than three times the width. Rounded indications are those in which the length is three times the width or less.

② Acceptance Standards.

Relevant indications shall be unacceptable when any of the following conditions exists:

a. Linear indications evaluated as crater cracks or star cracks exceed 5/32 inch (3.96 mm) in length.

b. Linear indications are evaluated as cracks other than crater cracks or star cracks.

c. Linear indications are evaluated as IF and exceed 1 inch (25.4 mm) in total length in a continuous 12-inch (304.8 mm) length of weld or 8 percent of the weld length.

d. Rounded indications shall be evaluated as follows:

i) Individual or scattered porosity (P) shall be unacceptable when any of the following conditions exists:

• The size of an individual pore exceeds 1/8 inch (3.17 mm).

• The size of an individual pore exceeds 25 percent of the thinner of the nominal wall thickness joined.

• The distribution of scattered porosity exceeds the concentration permitted by the porosity charts supplied by API for radiography testing.

ii) Cluster porosity (CP) that occurs in any pass except the finish pass shall comply with the criteria of (i). CP that occurs in the finish pass shall be unacceptable when any of the following conditions exists:

• The diameter of the cluster exceeds 1/2 inch (12.7 mm).

• The aggregate length of CP in any continuous 12 inch (304.8 mm) length of weld exceeds 1/2 inch (12.7 mm).

• An individual pore within a cluster exceeds 1/16 inch (1.59 mm) in size.

(5) ANSI/ASME B31.3, Chemical Plant and Petroleum Refinery Piping.

Acceptance Criteria:

Any crack or linear indication is unacceptable.

12. Personnel Qualification

(1) Personnel performing examinations to this procedure shall be qualified and certified in accordance with the requirements of ASNT document SNT-TC-1A (1992 edition), ISO 9712, PCN or Accepted level equivalent.

(2) Only individuals qualified to NDT Level II or individuals qualified to NDT Level I and working under the supervision of an NDT Level II may perform the examinations in accordance with this procedure.

(3) Evaluation of test results shall be by NDT Level II only.

13. Post cleaning

The inspection area shall be cleaned after inspection with solvent (unless otherwise specified) to remove any excess residues from the inspection process.

14. Demagnetisation

Demagnetisation shall be carried out where specifically requested by the client. The method to be employed shall be subject to agreement with the client.

15. Reporting

(1) A report shall be prepared detailing the results of the examination.

(2) The report shall contain sufficient information to enable a full assessment of quality and to ensure that any non-acceptable areas can be accurately located should repair be necessary.

(3) Any datum's used shall be unambiguous.

(4) Any restrictions to test shall be noted.

(5) A statement of acceptability against the acceptance standard should be made.

16. Referenced Documents
- ASME SECTION V ARTICLE 7
- ASME SECTION VIII, Division I
- ANSI/ASME B31.1
- ANSI/ASME B31.1
- ANSI/AWS D1.1
- ASTM E 709
- API 1104
- SNT-TC-1A of ASNT

03 Liquid Penetrant Testing

Item 3

Learning Objectives

1. Knowledge objectives
(1) To grasp the words, related terms and abbreviations about PT.
(2) To grasp the classification about PT system.
(3) To know the Instruments and Equipment of PT system.
(4) To know the testing procedure of PT system.

2. Competence objectives
(1) To be able to read and understand frequently used & complex sentence patterns, capitalized English materials and obtain key information quickly.
(2) To be able to communicate with English speakers about the topic freely.
(3) To be able to fill in the job cards in English.

3. Quality objectives
(1) To be able to self-study with the help of aviation dictionaries, the Internet and other resources.
(2) To do a good job of detection of safety protection.

3.1 Theory of Penetrant Testing

Liquid penetrant testing is a process used for testing and inspecting various objects for defects or flaws. It is a non-destructive type of testing process, which means that the object is not damaged or destroyed as it is tested. Many manufacturing and industrial facilities rely on liquid penetrant testing to confirm the safety and quality of their products.

During liquid penetrant testing, a dye is applied to the surface of an object. Through a process known as "capillary action", the dye quickly penetrates any flaws or cracks in the object's exterior. After a pre-determined period of time, the object is wiped clean, and all surface dye is removed. A developing powder is applied, which helps to brighten or expose any remaining dye that has settled into surface cracks or voids. The remaining dye indicates the location of any flaws or defects that must be corrected on the object.

There are two basic varieties of liquid penetrant testing used by manufacturers and inspectors. The first uses a visible, colored dye to indicate flaws, while the other relies on a fluorescent dye that is invisible to the naked eye. When the fluorescent dye is used, inspectors must examine the object under an ultraviolet (UV) light to view defects.

Liquid penetrant testing may be used on both metal and nonmetal objects, including ceramic, plastic and even glass. It is commonly used in manufacturing plants, to test things like cars, aeronautical equipment, machinery and consumer items. Nondestructive liquid penetrant testing is also an excellent method of inspecting building components without damaging or destroying the object during the process.

This testing process helps to reveal defects such as cracks, holes or voids that may not be detected using visual or other testing procedures. Liquid penetrant surface testing can also detect welding flaws in the seam between two different items. Finally, this test can be used to detect corrosion or other chemical processes that may have compromised the structural integrity of the object.

Liquid penetrant testing is known for its relatively low cost, especially compared to other materials testing techniques. The process is easy to perform and results are available quickly. Even inspectors who are relatively new to this testing process can easily interpret the results of this test in most cases.

While this test offers many benefits to users, it also has a number of limitations that must be considered. Liquid penetrant testing is only designed to find flaws on the surface of an object and will not detect problems deep below the surface. In addition, the surface of an object must be relatively smooth and free of dirt, grease and other substances that can skew the results of the test.

■ 〖Point 1〗 Basic Principle

This method of Testing works on the principle of capillary action. Liquid Penetrant is applied by immersion, spraying or brushing to the part to be inspected. The penetrant works its way into surface openings via capillary action. Excess surface penetrant is removed while allowing the penetrant in flaws to remain. A developer is applied which acts as a blotter to draw the penetrant from the flaws creating an indication on the surface of the part. This indication is either visible or fluorescent depending on the type of penetrant used.

The penetrant testing process (Fig. 3-1):

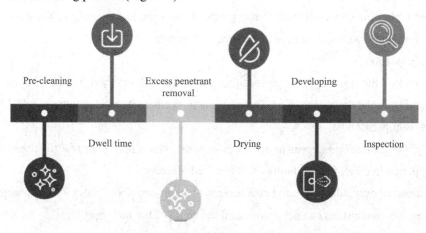

Fig. 3-1　Penetrant Testing Process

Step 1: Preliminary cleaning.

The components which are to be tested must be cleaned superficially so that the penetrant (Fig. 3-2) can penetrate any existing defects. Residues on the surface of the material, such as scale, slag and rust etc. must be removed by brushing, sanding, grinding and if necessary, by abrasive blasting. Care must be taken to ensure that the faults are not sealed by the cleaning process. The surface of the components must dry without residue.

Step 2: Penetrant process.

The penetrant can be applied by spraying, rinsing or immersing the components which are to be tested. It is essential to ensure that the surface is thoroughly wetted. The penetration time (also called dwell time) is strongly dependent on the surface and ambient temperatures. The dwell time is longer at low temperatures. The test temperature can range from −20 ℃ to +100 ℃.

Step 3: Excess penetrant removal.

The penetrant is removed from the surface by rinsing or spraying with water or a solvent-based cleaner. Beware of washing out: The penetrant must not be washed out of the cracks.

Step 4: Drying procedure.

After removing the excess penetrant with water or another cleaner, the surface should be dried with compressed air, a lint-free cloth or a suitable drying oven. If a cleaner is used which

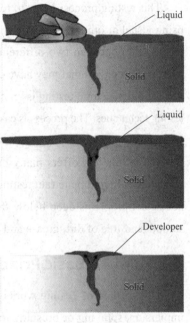

Fig. 3-2 Penetrant

dries by evaporation due to highly volatile components, the drying process can be omitted.

Step 5: Developing.

Immediately after drying, the developer is applied thinly and evenly. Aerosol spray cans or compressed air sprayers are particularly suitable.

The developing time depends on the temperature of the component's surface. The developing time is longer at low temperatures; at high temperatures it is shorter.

Step 6: Inspection.

At the end of the development period, the test surface is scanned for indications, so-called inhomogeneities, in the developer layer. The assessment needs to be carried out under UV light when using fluorescent penetrants.

Initially, the indications appear as red lines or dots, which may continue to spread during the developing period in case of larger faults—this is called bleeding.

The indication only allows limited conclusions to be drawn about the width or diameter of the fault opening. No conclusions can be drawn about the depth of the fault opening.

The advantage of liquid penetrant testing (LPI) offers over an unaided visual testing is that it makes defects easier to see for the inspector. There are basically two ways that a penetrant testing process makes flaws more easily seen. First, LPI produces a flaw indication that is much larger and easier for the eye to detect than the flaw itself. Many flaws are so small or narrow that they are undetectable by the unaided eye. Due to the physical features of the eye, there is a threshold below which objects cannot be resolved. This threshold of visual acuity is around 0.003 inch for a person with 20/20 vision.

The second way that LPI improves the detectability of a flaw is that it produces a flaw indication with a high level of contrast between the indication (Fig. 3–3) and the background also helping to make the indication more easily seen. When a visible dye penetrant testing is performed, the penetrant materials are formulated using a bright red dye that provides for a high level of contrast between the white developer. In other words, the developer serves as a high contrast background as well as a blotter to pull the trapped penetrant from the flaw. When a fluorescent penetrant testing is performed, the penetrant materials are formulated to glow brightly and to give off light at a wavelength that the eye is most sensitive to under dim lighting conditions.

Additional information on the human eye can be found by following the links below.

Fig. 3–3　Indications

〖Point 2〗 History

A very early surface testing technique involved the rubbing of carbon black on glazed pottery, whereby the carbon black would settle in surface cracks rendering them visible. Later, it became the practice in railway workshops to examine iron and steel components by the "oil and whiting" method. In this method, a heavy oil commonly available in railway workshops was diluted with kerosene in large tanks so that locomotive parts such as wheels could be submerged. After removal and careful cleaning, the surface was then coated with a fine suspension of chalk in alcohol so that a white surface layer was formed once the alcohol had evaporated. The object was then vibrated by being struck with a hammer, causing the residual oil in any surface cracks to seep out and stain the white coating. This method was in use from the latter part of the 19th century to approximately 1940, when the magnetic particle method was introduced and found to be more sensitive for ferromagnetic iron and steels.

A different (though related) method was introduced in the 1940s. The surface under examination was coated with a lacquer, and after drying, the sample was caused to vibrate by the tap of a hammer. The vibration causes the brittle lacquer layer to crack generally around surface defects. The brittle lacquer (stress coat) has been used primarily to show the distribution of stresses in a part and not for finding defects.

Many of these early developments were carried out by Magnaflux in Chicago, IL, USA in association with Switzer Bros., Cleveland, OH, USA. More effective penetrating oils containing highly visible (usually red) dyes were developed by Magnaflux to enhance flaw detection capability. This method, known as the visible or color contrast dye penetrant method, is still used quite extensively today. In 1942, Magnaflux introduced the Zyglo system of penetrant testing where fluorescent dyes were added to the liquid penetrant. These dyes would then fluoresce when exposed to ultraviolet light (sometimes referred to as "black light") rendering indications from cracks and other surface flaws more readily visible to inspectors.

〖Point 3〗 Advantages and Limitations

This method of Testing can be used on virtually any material and the results appear directly on the surface of the part. This method of testing is used predominantly on nonferrous materials (aluminum, titanium, magnesium, etc.).

The single largest limitation of this method is that it will only reveal flaws that are physically open to the surface. Therefore, surface preparation is critical to the effectiveness of this testing method. In certain instances, it is necessary to perform a pre-penetrant etch to remove smeared metal or excessive oxides that may have formed blocking the opening to flaws.

Liquid penetrant testing (LPI) is one of the most widely used non-destructive evaluation (NDE) methods. Its application scope in Fig. 3-4. Its popularity can be attributed to two main factors: its relative ease of use and its flexibility. LPI can be used to inspect almost any material provided that its surface is not extremely rough or porous. Materials that are commonly inspected using LPI include the following: metals (aluminum, copper, steel, titanium, etc.), glass, many ceramic materials, rubber, plastics.

Common PT Findings:			
√ Fatigue Cracks	√ Impact Fractures	√ Overload Fractures	√ Quench Cracks
√ Grinding Cracks	√ Laps	√ Porosity and Seams	√ Forging Defects
Common Materials Tested:			
√ Aluminum	√ Composites	√ Plastic	√ Stainless Steel
√ Ceramic Materials	√ Nickle	√ Rubber	√ Various Metals √ Glass
Common PT Applications:			
√ Engine Parts	√ Forged Parts	√ Pressure Bodies	√ Valve Components
√ Fan Blades	√ Inlet/Outlet Blowers	√ Stainless Steel	√ Weld Inspections
√ Fillet Welds	√ In-Service Parts	√ Turbine Engines	
Industries Using PT:			
√ Aerospace	√ Manufacturing	√ Pharmaceutical	√ Pull and Paper
√ Automotive	√ Ship Yards	√ Pipeline	√ Welding Fabrication
√ Construction	√ Petrochemical	√ Power Generation	

Fig. 3-4 Application

LPI offers flexibility in performing testings because it can be applied in a large variety of applications ranging from automotive spark plugs to critical aircraft components. Penetrant materials can be applied with a spray can or a cotton swab to inspect for flaws known to occur in a specific area or it can be applied by dipping or spraying to quickly inspect large areas. In the image above, visible dye penetrant is being locally applied to a highly loaded connecting point to check for fatigue cracking.

Penetrant testing systems have been developed to inspect some very large components. In the image shown right, DC-10 banjo fittings are being moved into a penetrant testing system at what used to be the Douglas Aircraft Company's Long Beach, California facility. These large machined aluminum forgings are used to support the number two engine in the tail of a DC-10 aircraft.

Liquid penetrant testing can only be used to inspect for flaws that break the surface of the sample. Some of these flaws (Fig. 3-5) are listed below: fatigue cracks, quench cracks, grinding cracks, overload and impact fractures, porosity, laps, seams, pin holes in welds, lack of fusion or braising along the edge of the bond line.

As mentioned above, one of the major limitations of a penetrant testing is that flaws must be open to the surface.

Here is one of our fluorescent dye penetrant test reports as an example.

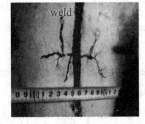

Fig. 3-5 Flaw

With fluorescent dye penetrant testing, minute material damage (such as cracks) becomes visible.

■ 〘Point 4〙 Capillary Action

Even if you've never heard of capillary action, it is still important in your life. Capillary action is important for moving water (and all of the things that are dissolved in it) around. It is defined as the movement of water within the spaces of a porous material due to the forces of adhesion, cohesion and surface tension.

Capillary action occurs because water is sticky, thanks to the forces of cohesion (water molecules like to stay close together) and adhesion (water molecules are attracted and stick to other substances). Adhesion of water to the walls of a vessel will cause an upward force on the liquid at the edges and result in a meniscus which turns upward. The surface tension acts to hold the surface intact. Capillary action occurs when the adhesion to the walls is stronger than the cohesive forces between the liquid molecules. The height to which capillary action will take water in a uniform circular tube (picture to right) is limited by surface tension and, of course, gravity.

Not only does water tend to stick together in a drop, it sticks to glass, cloth, organic tissues, soil, and luckily, to the fibers in a paper towel. Dip a paper towel into a glass of water and the water will "climb" onto the paper towel. In fact, it will keep going up the towel until the pull of gravity is too much for it to overcome.

Capillary action… in action! Without capillary action, the water level in all tubes would be the same. Smaller diameter tubes have more relative surface area inside the tube, allowing capillary action to pull water up higher than in the larger diameter tubes.

Capillary action (Fig. 3-6) is all around us every day.

If you dip a paper towel in water, you will see it "magically" climb up the towel, appearing to ignore gravity. You are seeing capillary action in action, and "climbing up" is about right—the water molecules climb up the towel and drag other water molecules along. (Obviously, Mona Lisa is a big fan of capillary action!)

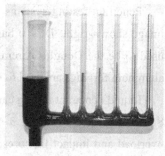

Fig. 3-6 Capillary Action

Plants and trees couldn't thrive without capillary action. Plants put down roots into the soil which are capable of carrying water from the soil up into the plant. Water, which contains dissolved nutrients, gets inside the roots and starts climbing up the plant tissue. Capillary action helps bring water up into the roots. But capillary action can only "pull" water up a small distance, after which it cannot overcome gravity. To get water up to all the branches and leaves, the forces of adhesion and cohesion go to work in the plant's xylem to move water to the furthest leaf.

Capillary action is also essential for the drainage of constantly produced tear fluid from the eye. Two tiny-diameter tubes, the lacrimal ducts, are present in the inner corner of the eyelid; these ducts secrete tears into the eye.

Maybe you've used a fountain pen…or maybe your parents or grandparents did. The ink moves from a reservoir in the body of the pen down to the tip and into the paper (which is composed of tiny paper fibers and air spaces between them) and not just turning into a blob. Of course gravity is responsible for the ink moving "downhill" to the pen tip, but capillary action is needed to keep the ink flowing onto the paper.

■ 〖Point 5〗 Wetting Action (Fig. 3-7)

Wetting refers to the study of how a liquid deposited on a solid (or liquid) substrate spreads out. Understanding wetting enables us to explain why water spreads readily on clean glass but no on a plastic sheet.

Example: Total wetting vs Non-wetting.

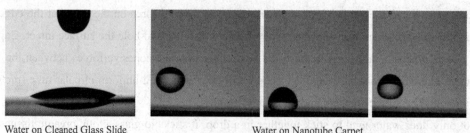

Water on Cleaned Glass Slide   Water on Nanotube Carpet

Fig. 3-7 Wetting Action

When a drop is placed down on very clean glass, it spreads completely. By contrast, the same drop deposited on a sheet of plastic remains stuck in its place. The conclusion is that there exist two regimes of wetting.

Wetting can be characterized into two types: total wetting, when the liquid has a strong affinity for the solid; and partial wetting, the opposite case.

Spreading Parameter

Spreading parameter, S distinguishes the two different regimes of wetting. It measures the difference between the surface energy (per unit area) of the substrate when dry and wet:

$$S = [E_{substrate}]_{dry} - E_{substrate\,wet} \text{ or } S = (\gamma_{liquid} + \gamma_{solid\text{-}liquid})$$

$S > 0$: Total wetting

If the parameter S is positive, the liquid spreads completely in order to lower its surface energy. Condition favorable for this condition is a high value of γ solid (high energy surfaces like glass, clean silicon) and a lower value of γ liquid (ethanol, toluene).

$S < 0$: Partial wetting

The drop does not spread but, instead, forms at equilibrium aspherical cap resting on the substrate with a contact angle θ. A liquid is said to be "mostly wetting" when $\theta < 90°$, and "mostly non-wetting" when $\theta > 90°$ (Fig. 3-8).

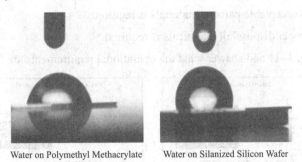

Fig. 3-8 Partially Wetting and Partially Non-wetting

When the solid has a high affinity for water—in which case it is called hydrophilic (high energy e.g. glass) —water spreads. In the opposite case of hydrophobic (low energy, e.g. Teflon) surfaces, water does not spread but, instead, forms at equilibrium a spherical cap resting on the substrate with a "contact angle" (Fig. 3-9).

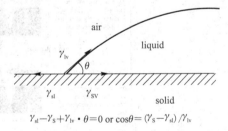

$\gamma_{sl} - \gamma_s + \gamma_{lv} \cdot \theta = 0$ or $\cos\theta = (\gamma_s - \gamma_{sl})/\gamma_{lv}$

Fig. 3-9 Contact Angle

Put into Practice

1. Analyze which wetting methods are shown in Figures A, B, and C (Fig. 3-10).

2. Read the following essay to analyze the advantages and disadvantages of penetration detection?

PT advantages:

√ High sensitivity to small surface discontinuities.

√ Suitable for large range of materials: metallic and non-

Fig. 3-10 Wetting Methods

metallic, magnetic and non-magnetic and conductive and non-conductive.

√ Large surface areas and large volumes parts can be inspected rapidly and at low cost.

√ Parts with complex geometry can be tested without changing the inspection procedure.

√ Indications are produced directly on the surface of the part.

√ Aerosol spray cans make penetrant materials very portable.

√ Required materials and equipment are relatively inexpensive.

PT limitations:

√ Only surface breaking defects can be detected.

√ Only materials with relatively nonporous surface can be inspected.

√ Precleaning is critical.

√ Metal smearing from machining, grinding, and grit or vapor blasting must be removed prior to inspection.

√ Surface finish and roughness can affect inspection sensitivity.

√ Multiple process operations must be performed with waiting intervals.

√ Post cleaning of acceptable parts or materials is required.

√ Handling and proper disposal of chemicals is required.

3. Analyze the Fig. 3–11 and answer what the operational requirements are for each step.

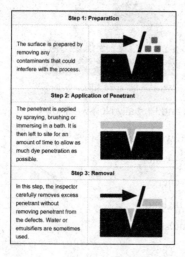 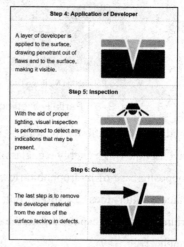

Fig. 3–11　Operation Steps

3.2　Method of PT

The basic principle of liquid penetrant testing (PT) is capillary action, which allows the penetrant to enter in the opening of the defect, remain there when the liquid is removed from the material surface and then re-emerge on the surface on application of a developer, which has a capillary action similar to blotting paper. The term penetrant material includes all penetrants, solvents or cleaning agents that are used in this examination process (Fig. 3-12). A penetrant material has the capacity to enter the crevices opening on the surface of a material. Fluorescent or visible penetrant with color contrast are used with one of the following three penetrant processes, namely, water washable, post-emulsifying and solvent removable. The combination of fluorescent or visible penetrant with the three processes results in six possible liquid penetrant techniques. In the color contrast penetrant process, the developer forms a reasonably uniform white coating. The fluorescent penetrant process is similar to the color contrast process except that the examination is performed using ultraviolet light, which is also called black light.

Penetrant		Cleaner		Developer	
Type	Designation	Method	Designation	Form	Designation
I	Fluorescent penetrant	A	Water	a	Dry
II	colour contrast penetrant	B	Lipophilic emulsifier 1. Oil-based emulsifier 2. Rinsing with running Water	b	Water-soluble
				c	Water-suspendable
III	Dual purpose (fluorescent colour contrast penetrant)	C	Solvent (liquid)	d	Solvent-based (non-aqueous wet)
		D	Hydrophilic emulsifier 1. Optional prerinse (water) 2. Emulsifier (water-diluted) 3. Final rinse (water)	e	Water or solvent based for special application (e.g.peelable developer)
		E	Water and solvent		
Note: For specific cases, it is necessary to use penetrant testing products complying with particular requirements. With regards to flammability, sulfur, halogen and sodium co ntent and other contaminants, see prEN 571-2.					

Fig. 3-12 Classification

■ 〖Point 1〗 Classification of Penetrant Testing Materials

The penetrant materials used today are much more sophisticated than the kerosene and whiting first used by railroad inspectors near the turn of the 20th century. Today's penetrants are carefully formulated to produce the level of sensitivity desired by the inspector. To perform well, a penetrant must possess a number of important characteristics. A penetrant must:

• Spread easily over the surface of the material being inspected to provide complete and even coverage.

• Be drawn into surface breaking defects by capillary action.

• Remain in the defect but remove easily from the surface of the part.

• Remain fluid so it can be drawn back to the surface of the part through the drying and developing steps.

- Be highly visible or fluoresce brightly to produce easy to see indications.
- Not be harmful to the material being tested or the inspector.

All penetrant materials do not perform the same and are not designed to perform the same. Penetrant manufactures have developed different formulations to address a variety of testing applications. Some applications call for the detection of the smallest defects possible and have smooth surfaces where the penetrant is easy to remove. In other applications, the rejectable defect size may be larger and a penetrant formulated to find larger flaws can be used. The penetrants that are used to detect the smallest defect will also produce the largest amount of irrelevant indications.

Penetrant materials are classified in the various industry and government specifications by their physical characteristics and their performance. Aerospace Material Specification (AMS) 2644, Testing Material, Penetrant, is now the primary specification used in the USA to control penetrant materials. Historically, Military Standard 25135, Testing Materials, Penetrants, has been the primary document for specifying penetrants but this document is slowly being phased out and replaced by AMS 2644. Other specifications such as ASTM 1417, Standard Practice for Liquid Penetrant Examinations, may also contain information on the classification of penetrant materials but they are generally referred back to MIL-I-25135 or AMS 2644.

Penetrant materials come in two basic types. These types are listed below (Fig. 3–13, Fig. 3–14):

Fig. 3–13   Type 1—Fluorescent Penetrants

Fig. 3–14   Type 2—Visible Penetrants

Fluorescent penetrants contain a dye or several dyes that fluoresce when exposed to ultraviolet radiation. Visible penetrants contain a red dye that provides high contrast against the white developer background. Fluorescent penetrant systems are more sensitive than visible penetrant systems because the eye is drawn to the glow of the fluorescing indication. However, visible penetrants do not require a darkened area and an ultraviolet light in order to make an Testing. Visible penetrants are also less vulnerable to contamination from things such as cleaning fluid that can significantly reduce the strength of a fluorescent indication.

〖Point 2〗 Classification of remove the excess penetrant

Penetrants are then classified by the method used to remove the excess penetrant from the part. The four methods are listed below:

- Method A—Water Washable
- Method B—Post-Emulsifiable, Lipophilic
- Method C—Solvent Removable
- Method D—Post-Emulsifiable, Hydrophilic

Water washable (Method A) penetrants can be removed from the part by rinsing with water alone. These penetrants contain an emulsifying agent (detergent) that makes it possible to wash the penetrant from the part surface with water alone. Water washable penetrants are sometimes referred to as self-emulsifying systems.

Post-emulsifiable penetrants come in two varieties, lipophilic and hydrophilic. In post-emulsifiers, lipophilic systems (Method B), the penetrant is oil soluble and interacts with the oil-based emulsifier to make removal possible.

Post-emulsifiable, hydrophilic systems (Method D) (Fig. 3-15), use an emulsifier that is a water soluble detergent which lifts the excess penetrant from the surface of the part with a water wash. Solvent removable penetrants require the use of a solvent to remove the penetrant from the part.

Fig. 3-15 Post-emulsifiable, Hydrophilic Systems

〖Point 3〗 Classification of Sensitivity Levels

Penetrants are then classified based on the strength or detectability of the indication that is produced for a number of very small and tight fatigue cracks. The five sensitivity levels are shown below:

- Level ½—Ultra Low Sensitivity
- Level 1—Low Sensitivity
- Level 2—Medium Sensitivity
- Level 3—High Sensitivity
- Level 4—Ultra-High Sensitivity

The major US government and industry specifications currently rely on the US Air Force

Materials Laboratory at Wright-Patterson Air Force Base to classify penetrants into one of the five sensitivity levels (Table 3–1). This procedure uses titanium and Inconel specimens with small surface cracks produced in low cycle fatigue bending to classify penetrant systems. The brightness of the indication produced is measured using a photometer. The sensitivity levels and the test procedure used can be found in Military Specification MIL-I-25135 and Aerospace Material Specification 2644, Penetrant Testing Materials.

Table 3–1 Sensitivity Levels

Type Ⅰ—Fluorescent	Method A—Water Washable	Sensitivity Levels 1-4
Type Ⅱ—Visible dye	Method B—Post Emulsified (lipophilic)	
	Method C—Solvent Removable	
	Method D—Post Emulsified (hydrophilic)	

■ 〖Point 4〗 Classification of Development

An interesting note about the sensitivity levels is that only four levels were originally planned. However, when some penetrants were judged to have sensitivities significantly less than most others in the level 1 category, the ½ level was created. An excellent historical summary of the development of test specimens for evaluating the performance of penetrant materials can be found in the following reference. Developers (Fig. 3–16) shall be of the following forms:

Form a—Dry powder

Form b—Water soluble

Form c—Water suspendible

Form d—Nonaqueous Type 1 Fluorescent (solvent based)

Form e—Nonaqueous Type 2 Visible Dye (solvent based)

Form f—Special application

Fig. 3–16 Developer

Dry powder developer is generally considered to be the least sensitive but it is inexpensive to use and easy to apply. Dry developers are white, fluffy powders that can be applied to a thoroughly dry surface in a number of ways. The developer can be applied by dipping parts in a container of developer, or by using a puffer to dust parts with the developer. Parts can also be placed in a dust cabinet where the developer is blown around and allowed to settle on the part. Electrostatic powder spray guns are also available to apply the developer. The goal is to allow the developer to come in contact with the whole testing area.

Unless the part is electrostatically charged, the powder will only adhere to areas where trapped penetrant has wet the surface of the part. The penetrant will try to wet the surface of the penetrant particle and fill the voids between the particles, which brings more penetrant to the surface of the part where it can be seen. Since dry powder (Fig. 3–17) developers only stick to the area where penetrant is present, the dry developer does not provide a uniform white background as the other forms of developers do. Having a

Fig. 3–17 Dry Powder

uniform light background is very important for a visible testing to be effective and since dry developers do not provide one, they are seldom used for visible testings. When a dry developer is used, indications tend to stay bright and sharp since the penetrant has a limited amount of room to spread.

As the name implies, water soluble developers consist of a group of chemicals that are dissolved in water and form a developer layer when the water is evaporated away. The best method for applying water soluble developers is by spraying it on the part. The part can be wet or dry. Dipping, pouring or brushing the solution on to the surface is sometimes used but these methods are less desirable. Aqueous developers contain wetting agents that cause the solution to function much like dilute hydrophilic emulsifier and can lead to additional removal of entrapped penetrant. Drying is achieved by placing the wet but well drained part in a recirculating, warm air dryer with the temperature held between 70 °F and 75 °F. If the parts are not dried quickly, the indications will be blurred and indistinct. Properly developed parts will have an even, pale white coating over the entire surface.

Water suspendable developers (Fig. 3–18) consist of insoluble developer particles suspended in water. Water suspendable developers require frequent stirring or agitation to keep the particles from settling out of suspension. Water suspendable developers are applied to parts in the same manner as water soluble developers. Parts coated with a water suspendable

Fig. 3–18 Water Suspendable Developers

89

developer must be forced dried just as parts coated with a water soluble developer are forced dried. The surface of a part coated with a water suspendable developer will have a slightly translucent white coating.

Nonaqueous developers (Fig.3-19) suspend the developer in a volatile solvent and are typically applied with a spray gun. Nonaqueous developers are commonly distributed in aerosol spray cans for portability. The solvent tends to pull penetrant from the indications by solvent action. Since the solvent is highly volatile, forced drying is not required. A nonaqueous developer should be applied to a thoroughly dried part to form a slightly translucent white coating.

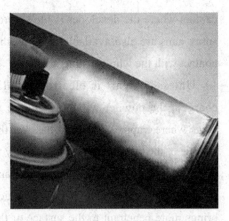

Fig. 3-19 Nonaqueous Developers

Plastic or lacquer developers are special developers that are primarily used when a permanent record of the testing is required.

〖Point 5〗 Classification of Solvent Removers

Solvent Removers shall be of the following classes:

Class 1 Halogenated

Class 2 Nonhalogenated

Class 3 Special application

Put into Practice

1. The following diagram is analyzed to show what visible agents are used for each penetration detection method corresponding to the sensitivity level (Table 3-2) and what exactly is the method in which the image agent is applied.

Table 3-2 Sensitivity Level

Ranking	Developer Form	Method of Application
1	Nonaqueous, Wet Solvent	Spray
2	Plastic Film	Spray
3	Water-Soluble	Spray
4	Water-Suspendable	Spray
5	Water-Soluble	Immersion
6	Water-Suspendable	Immersion
7	Dry	Dust Cloud (Electrostatic)
8	Dry	Fluidized Bed
9	Dry	Dust Cloud (Air Agitation)
10	Dry	Immersion (Dip)

2. Look at Fig. 3–20 to show what are the advantages and disadvantages of each image?

Developer	Advantages	Disadvantages
Dry	Indications tend to remain brighter and more distinct over time Easy to apply	Does not form contrast background so cannot be used with visible systems Difficult to assure entire part surface has been coated
Soluble	Ease of coating entire part White coating for good contrast can be produced which work well for both visible and fluorescent systems	Coating is translucent and provides poor contrast (not recommended for visual systems) Indications for water washable systems are dim and blurred
Suspendable	Ease of coating entire part Indications are bright and sharp White coating for good contrast can be produced which work well for both visible and fluorescent systems	Indications weaken and become diffused after time
Nonaqueous	Very portable Easy to apply to readily accessible surfaces White coating for good contrast can be produced which work well for both visible and fluorescent systems Indications show-up rapidly and are well defined Provides highest sensitivity	Difficult to apply evenly to all surfaces More difficult to clean part after inspection

Fig. 3–20　The Advantages and Disadvantages

3.3　Instruments and Equipment of Penetrant Testing

Penetrant systems range from simple portable kits to large, complex in-line test systems. Standard manual tank line generally includes: pre-cleaning tanks, penetrant immersion tank, manual rinsing tanks, emulsifier tank, drying tanks, developing tanks and testing booth. This is mostly common for aerospace and medical industry.

〖Point 1〗 Penetrant Kit (Fig. 3-21)

The kits contain pressurized cans of the penetrant, cleaner, remover, solvent, developer and in some cases, brushes, swabs and cloths. A larger fluorescent penetrant kit will include a black light. These kits are used when examinations are to be conducted in remote areas, in the field or for a small area of a test surface. In contrast to these portable.

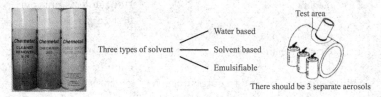

Fig. 3-21 Penetrant Kit

〖Point 2〗 Standard Manual Tank Line

Penetrant kits, there are a number of diverse stationary-type systems. These range from a manually operated penetrant line with a number of tanks, to very expensive automated lines, in which most steps in the process are performed automatically. The penetrant lines can be very simple, as illustrated in Fig. 3-22.

In this particular system, there is a tank for the penetrant, a tank for the water rinse, a drying oven and a developer station. The final station is the examination area, which includes a black light. This manually operated system is a typical small water-removable penetrant line. The steps in the testing process would be: cleaning of the parts, application of the penetrant, removal of the penetrant with a water spray, drying, application of the developer, and finally, testing. This entire process is covered in much greater detail in Section V, Techniques.

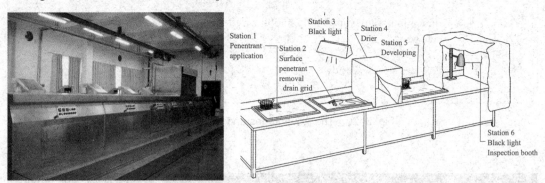

Fig. 3-22 Tipical Fluorescent Pemetrant Line Arrangement

If a poste-mulsifiable penetrant is to be used, the manually operated line will require an additional tank. This tank will contain an emulsifier that will render the surface penetrant.

〖Point 3〗 FPI Spray Booth

FPI spray booth (Fig. 3-23) includes electrostatic penetrant spray booth, rinsing booth, drying booth, developers spray booth and testing booth. It is mainly used by big parts testing for aerospace industry.

Fig. 3-23　FPI Spray Booth

■〖Point 4〗 Automatic Tank Lines

Automatic tank line includes all tanks for FPI, but also with automatic crane system and PLC control. It is mainly used for automotive industry with very large production.

The automatic penetrant lines in use today vary from small, rather simple systems to very large complex lines that are computer controlled. Fig. 3-24 illustrates a large automatic penetrant line.

Fig. 3-24　Automatic Tank Line

Although the steps in an automated penetrant system have been somewhat mechanized, it is interesting to note that the examinations still must be conducted by inspectors who have been trained and are qualified in the process. The arrangement of these large automated penetrant lines vary with different layouts to permit the most flexibility from the standpoint of processing the parts. Normally, the systems will be arranged in a straight line; however, U shape or other configuration may be used to provide more effective use of floor space.

■〖Point 5〗 Other Accessories

The black light is an essential accessory for fluorescent penetrant testing. Black lights used in penetrant testing typically produce wavelengths in the range of 315 to 400 nm (3 150–4 000 Angstrom units) and utilize mercury vapor bulbs of the sealed-reflector type. These lights are provided with a "Woods" filter, which eliminates the undesirable longer wavelengths. Black light intensity requirements will range from 800 to 1 500 microwatts per square centimeter (W/cm^2) at the test surface. Specific requirements will vary, depending upon the code or specification(s) being used. Recent developments in black light technology provide lights that can produce intensities up to 4 800 W/cm^2 at light intensity meters are used to measure both white light intensities when visible PT is used and black light intensities for fluorescent penetrant techniques. This measurement is necessary to verify code compliance and to assure that there is no serious degradation of the lights. Some meters are designed to measure both white light and black light intensities.

Test panels, including comparator blocks, controlled cracked panels, tapered plated panels and others for specific industries such as the TAM panel (typically used in aerospace), are employed to control the various attributes of the PT system. They also provide a means for monitoring the materials and the process. This is discussed in Quality Control Considerations.

System performance checks involve processing a test specimen with known defects to determine if the process will reveal discontinuities of the size required. The specimen must be processed following the same procedure used to process production parts. A system performance check is typically required daily, at the reactivation of a system after maintenance or repairs, or any time the system is suspected of being out of control. As with penetrant testings in general, results are directly dependent on the skill of the operator and, therefore, each operator should process a panel.

The ideal specimen is a production item that has natural defects of the minimum acceptable size. Some specification delineates the type and size of the defects that must be present in the specimen and detected. Surface finish is will affect washability so the check specimen should have the same surface finish as the production parts being processed. If penetrant systems with different sensitivity levels are being used, there should be a separate specimen for each system.

There are some universal test specimens that can be used if a standard part is not available. The most commonly used test specimen is the TAM or PSM panel (Fig. 3–25). These panel are usually made of stainless steel that has been chrome plated on one half and surfaced finished on the other half to produced the desired roughness. The chrome plated section is impacted from the back side to produce a starburst set of cracks in the chrome. There are five impacted areas to produce range of crack sizes. Each panel has a characteristic "signature" and variances in that signature are indications of process variance. Panel patterns as well as brightness are indicators of process consistency or variance.

Care of system performance check specimens is critical. Specimens should be handled carefully to avoid damage. They should be cleaned thoroughly between uses and storage in a solvent is generally recommended. Before processing a specimen, it should be inspected under UV light to make sure that it is clean and not already producing an indication.

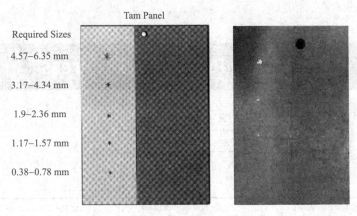

Fig. 3-25 Comparator Blocks

In summary, the equipment used will be greatly influenced by the size, shape and quantity of products that are to be examined. If there are large quantities involved on a continuing basis, the use of an automated system may be appropriate, whereas with small quantities of parts, the use of penetrant kits may be more suitable. The size and configuration of the part will also influence the type of penetrants that will be most appropriate.

Put into Practice

1. The names and effects of the penetration detection parts in Fig. 3-26 are pointed out.

A typical small sized penetrant system employing post-emulsified fluorescent penetrant with dry powder developer.

| Penetrant application station | Emulsifier application station | Drain and wash station Water pressure and temperature controlled | U.V Lamp | Drying oven. Temperature controlled | Developer application station Spray gun | U.V Lamp Inspection station |

Fig. 3-26 Typical Penetrant Systems

2. Point out the type, material and function of the test block used for penetration below (Fig. 3-27).

Description	Product
Pressure code, section V & III, MIL+25135 and AMS-2644. The cracked aluminum block is made from SB-211 Type 2024 aluminum, rolled 3/8 inch thick with dimensions of 2×3 inches. A notch separates the block's two sides to facilitate side-by-side comparisons.	Cracked Aluminum Blocks

Fig. 3-27 Test Blocks

3.4 The Penetrant Progress

The penetrant progress as shown in Fig. 3-28.

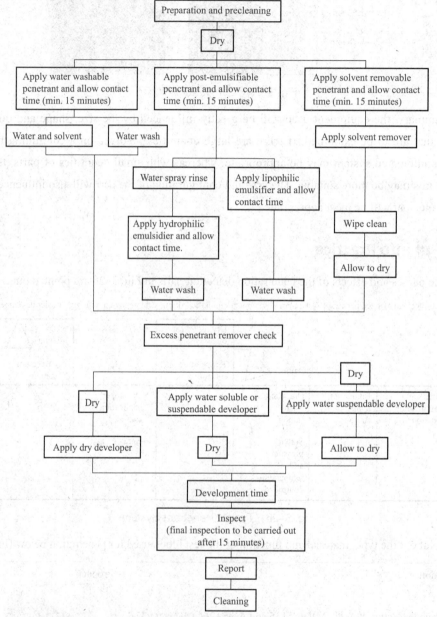

Fig. 3-28 The Penetrant Progress

■ 〖Point 1〗 Preclearing

One of the most critical steps in the penetrant testing process is preparing the part for testing. All coatings, such as paints, varnishes, plating and heavy oxides must be removed to ensure that defects are open to the surface of the part. If the parts have been machined, sanded or blasted prior to the penetrant testing, it is possible that a thin layer of metal may have smeared across the surface and closed off defects. It is even possible for metal smearing to occur as a result of cleaning operations such as grit or vapor blasting. This layer of metal smearing must be removed before testing.

Preclearing after addressing the prerequisites, it is necessary to remove all contaminants from the surface, after the surface has been cleaned, all evidence of any residues that may remain. After preclearing, it is essential that the drycleaners evaporate and that the test surface be totally dry prior to application of the penetrant. This will prevent contamination or dilution of the penetrant in the event that it interacts and becomes mixed with the recliner.

■ 〖Point 2〗 Penetrant Application

The penetrant can be applied to the surface of the test part in virtually any effective manner, including brushing, dipping the part into the penetrant, immersion, spraying or just pouring it on the surface (Fig. 3-29).

Fig. 3-29 Penetrant Applied Method

Fig. 3-30 shows a water-removable fluorescent penetrant being applied with an electrostatic sprayer. The key is to assure that the area of interest is effectively wetted and that the penetrant liquid does not dry during the penetration or dwell time, which is the period of time from when the penetrant is applied to the surface until it is removed. The codes and specifications give detailed dwell times that must be followed. It is quite common to have a dwell time of 10 to 15 minutes for many applications (Table 3-3).

Parts are solvent washed and immersed in a fluorescent dye

Liquid fluorescent penetrant inspetion is used for testing aerospace aluminum and other non-ferrous metals

Fig. 3-30 Penetrant Application

Table 3-3 Penetrant Time

Material	Form	Type of Discontinuity	Water-washable Penetration time*
Aluminium	Castings	Porosity, Cold Shuts	5 to 15 minutes
Aluminium	Extrusions, Forgings	Laps	NR**
Aluminium	Welds	Lack of Fusion, Porosity	30
Aluminium	All	Cracks, Fatigue Cracks	30 , not recommended for fatigue crack
Magnesium	Castings	Porosity, Cold Shuts	15
Magnesium	Extrusions, Forgings	Laps	not recommended
Magnesium	Welds	Lack of Fusion, Porosity	30
Magnesium	All	Cracks, Fatigue Cracks	30, not recommended for fatigue crack
Steel	Castings	Porosity, Cold Shuts	30
Steel	Extrusions, Forgings	Laps	not recommended
Steel	Wekis	Lack of Fusion, Porosity	60

〖Point 3〗 Penetrant Removal

Penetrant Removal in this step the excess surface penetrant is removed from the test specimen surface; the method of removal depends on the type of penetrant that is being used. There are three techniques for excess surface penetrant removal: water, emulsifiers and solvents.

Fig. 3-31 illustrates excess surface visible contrast penetrant being removed with a solvent dampened cloth. Removal of fluorescent penetrants is usually accomplished under a black light. This provides a means of assuring complete removal of the excess surface penetrant while minimizing the possibility of over removal.

Removal of excess surface penetrant

Examination is performed under UV light. Cracks and other surface irregularities become visible.

Fig. 3-31 Excess Surface Visible Contrast Penetrant

Method C, Solvent Removable, is used primarily for inspecting small localized areas. This method requires hand wiping the surface with a cloth moistened with the solvent remover, and is, therefore, too labor intensive for most production situations. Of the three production penetrant testing methods, Method A, water-washable, is the most economical to apply. Water-washable or self-emulsifiable penetrants contain an emulsifier (Fig. 3–32) as an integral part of the formulation. The excess penetrant may be removed from the object surface with a simple water rinse. These materials have the property of forming relatively viscous gels upon contact with water, which results in the formation of gel-like plugs in surface openings. While they are completely soluble in water, given enough contact time, the plugs offer a brief period of protection against rapid wash removal. Thus, water-washable penetrant systems provide ease of use and a high level of sensitivity.

Fig. 3–32 Emulsifier

When removal of the penetrant from the defect due to over-washing of the part is a concern, a post-emulsifiable penetrant system can be used. Post-emulsifiable penetrants require a separate emulsifier to breakdown the penetrant and make it water washable. The part is usually immersed in the emulsifier but hydrophilic emulsifiers may also be sprayed on the object. Spray application is not recommended for lipophilic emulsifiers because it can result in non-uniform emulsification if not properly applied. Brushing the emulsifier on to the part is not recommended either because the bristles of the brush may force emulsifier into discontinuities, causing the entrapped penetrant to be removed. The emulsifier is allowed sufficient time to react with the penetrant on the surface of the part but not given time to make its way into defects to react with the trapped penetrant. The penetrant that has reacted with the emulsifier is easily cleaned away. Controlling the reaction time is of essential importance when using a post-emulsifiable system. If the emulsification time is too short, an excessive amount of penetrant will be left on the surface, leading to high background levels. If the emulsification time is too long, the emulsifier will react with the penetrant entrapped in discontinuities, making it possible to deplete the amount needed to form an indication.

The hydrophilic post-emulsifiable method (Method D) (Fig. 3–33) is more sensitive than the lipophilic post-emulsifiable method (Method B) (Fig. 3–34). Since these methods are generally only used when very high sensitivity is needed, the hydrophilic method renders the lipophilic method virtually obsolete. The major advantage of hydrophilic emulsifiers is that they are less sensitive to variation in the contact and removal time. While emulsification time should be controlled as closely as possible, a variation of one minute or more in the contact time will have little effect on flaw detectability when a hydrophilic emulsifier is used. On the contrary, a variation of as little as 15 to 30 seconds can have a significant effect when a lipophilic system is used. Using an emulsifier involves adding a couple of steps to the penetrant process, slightly increases the cost of an testing. When using an emulsifier, the penetrant process includes the following steps: pre-clean part, apply penetrant and allow to dwell, pre-rinse to remove first layer of penetrant, apply hydrophilic emulsifier and allow contact for specified time, rinse to remove excess penetrant, dry part, apply developer and allow part to develop, inspect.

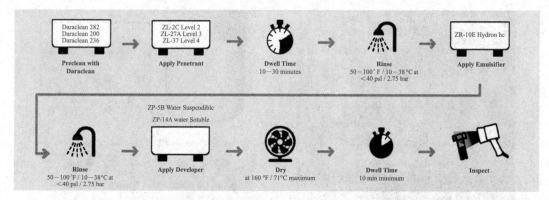

Fig. 3–33　Method D: The Hydrophilic Post-emulsifiable Fluorescent

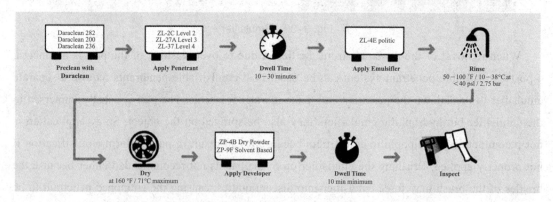

Fig. 3–34　Method B: The Lipophilic Post-emulsifiable Fluorescent

■ 〖Point 4〗 Application of Developer

The type of developer to be used will be specified in the penetrant procedure. As mentioned above, the four types of developers are dry, nonaqueous, aqueous suspendable and aqueous soluble. The entire test surface or area of interest must be properly developed, although there are rare applications where developers are not used. A nonaqueous developer is applied by spraying. It must be applied in a thin, uniform coating. Thick layers of developer, whether nonaqueous, dry, or aqueous, can tend to mask a discontinuity bleed-out, especially if that discontinuity is small and tight.

Development Time

The developer must be given ample time to draw the entrapped penetrant from the discontinuity out to the test surface. Many codes and specifications will require a development time from 7 to 30 minutes and, in some cases, as long as 60 minutes. Development is defined as the time it takes from the application of lists (Fig. 3-35) the main advantages and disadvantages of the various developer types. List in the characterizing and to determine the extent of the indication (s) .

The Developer Application Type	Post-emulsifiable Penetrant			Water-washable Penetrant		
Developer	Fatigue Crack	Porosity	Stress Corrosion Crack	Fatigue Crack	Porosity	Stress Corrosion Crack
None	20	20	20	5	10	20
Nonaqueous Wet	10	20	20	5	10	5
Dry	5	20	20	5	10	5
Aqueous Wet	5	10	20	5	20	*

Fig. 3-35   Developer Time

〖Point 5〗 Interpretation

Interpretation upon completion of the development time, the indications from discontinuities or other sources that have formed must be interpreted. A visible contrast penetrant bleed out is illustrated in Fig.3-36 fluorescent penetrant indications are shown in Table 3-4. Bleed Fig.3-36 outs are interpreted based primarily on their size, shape and intensity.

Table 3-4   Acceptance Levels for Indication

Type of indication	Acceptance level		
	1	2	3
Linear indication l=length of indication	$l \leqslant 2$ mm	$l \leqslant 4$ mm	$l \leqslant 8$ mm
Non-linear indication d=major axis dimension	$d \leqslant 4$ mm	$d \leqslant 6$ mm	$d \leqslant 8$ mm
Acceptance levels 2 and 3 may be specified with suffix "x" which denotes that all linear indications detected shall be evaluated to level 1. However, the probability of detection of indications smaller than those denoted by the original acceptance level can be low. Linear defect such like crack, lack of fusion and lack of penetration is not acceptable regardless of length.			

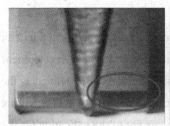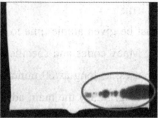

Fig. 3-36　Visible Contrast Penetrant

■〖Point 6〗 Postcleaning

After the part has been evaluated and the report completed, all traces of any remaining penetrant and developer must be thoroughly removed from the test surface prior to it being placed into service or returned for further processing.

■〖Point 7〗 Process Control of Temperature

The temperature of the penetrant materials and the part being inspected can have an effect on the results. Temperatures from 27 to 49 ℃ (80 to 120 °F) are reported in the literature to produce optimal results. Many specifications allow testing in the range of 4 to 52 ℃ (40 to 125 °F). A tip to remember is that surfaces that can be touched for an extended period of time without burning the skin are generally below 52 ℃ (125 °F).

Since the surface tension of most materials decrease as the temperature increases, raising the temperature of the penetrant will increase the wetting of the surface and the capillary forces. Of course, the converse is also true, so lowering the temperature will have a negative effect on the flow characteristics. Raising the temperature will also raise the speed of evaporation of penetrants, which can have a positive or negative effect on sensitivity. The impact will be positive if the evaporation serves to increase the dye concentration of the penetrant trapped in a flaw up to the concentration quenching point and not beyond. Higher temperatures and more rapid evaporation will have a negative effect if the dye concentration exceeds the concentration quenching point, or the flow characteristics are changed to the point where the penetrant does not readily flow.

The method of processing a hot part was once commonly employed. Parts were either heated or processed hot off the production line. In its day, this served to increase testing sensitivity by increasing the viscosity of the penetrant. However, the penetrant materials used today have 1/3 to 1/2 the viscosity of the penetrants on the market in the 1960s and 1970s. Heating the part prior to testing is no longer necessary and no longer recommended.

■〖Point 8〗 Quality Control of Penetrant

The quality of a penetrant testing is highly dependent on the quality of the penetrant materials used. Only products meeting the requirements of an industry specification, such as AMS 2644, should be used. Deterioration of new penetrants primarily results from aging and contamination. Virtually all organic dyes deteriorate over time, resulting in a loss of color or fluorescent response, but deterioration can be slowed with proper storage. When possible, keep the materials in a closed container and protect from freezing and exposure to high heat. Freezing can cause separation to occur and exposure to high temperature for a long period of time can affect the brightness of the dyes.

Contamination can occur during storage and use. Of course, open tank systems are much more susceptible to contamination than are spray systems. Contamination by another liquid will change the surface tension and contact angle of the solution. Water is the most common contaminant. Water-washable penetrants have a definite tolerance limit for water, and above this limit they do not function properly. Cloudiness and viscosity both increase with increasing water content. In self-emulsifiable penetrants, water contamination can produce a gel break or emulsion inversion when the water concentration becomes high enough. The formation of the gel is an important feature during the washing processes, but must be avoided until that stage in the process. Data indicates that the water contamination must be significant (greater than 10%) for gel formation to occur. Most specifications limit water contamination to around 5% to be conservative. Water does not readily mix with the oily solution of lipophilic post-emulsifiable systems and it generally settles to the bottom of the tank. However, the testing of parts that travel to the bottom of the tank and encounter the water could be adversely affected.

Most other common contaminates, such as cleaning solvents, oils, acids, caustics and chromates, must be present in significant quantities to affect the performance of the penetrant. Organic contaminants can dilute the dye and absorb the ultraviolet radiation before it reaches the dye, and also change the viscosity. Acids, caustics and chromates cause the loss of fluorescence in water-soluble penetrants.

Regular checks must be performed to ensure that the material performance has not degraded. When the penetrant is first received from the manufacturer, a sample of the fresh solution should be collected and stored as a standard for future comparison. The standard specimen should be stored in a sealed, opaque glass or metal container. Penetrants that are in-use should be compared regularly to the standard specimen to detect changes in color, odor and consistency. When using fluorescent penetrants, a brightness comparison per the requirements of ASTM E1417 is also often required. This check involves placing a drop of the standard and the in-use penetrants on a piece of Whatman #4 filter paper and making a side by side comparison of the brightness of the two spots under UV light.

Additionally, the water content of water washable penetrants must be checked regularly. Water-based, water-washable penetrants are checked with a refractometer. The rejection criteria are different for different penetrants, so the requirements of the qualifying specification or the manufacturer's instructions must be consulted. Non-water-based, water-washable penetrants are checked using the procedure specified in ASTM D95 or ASTM E1417.

Put into Practice

1. The following test procedures (Fig. 3–37) are classified as washing method, solvent removal method, or hydro-emulsification method and oil-based emulsification.

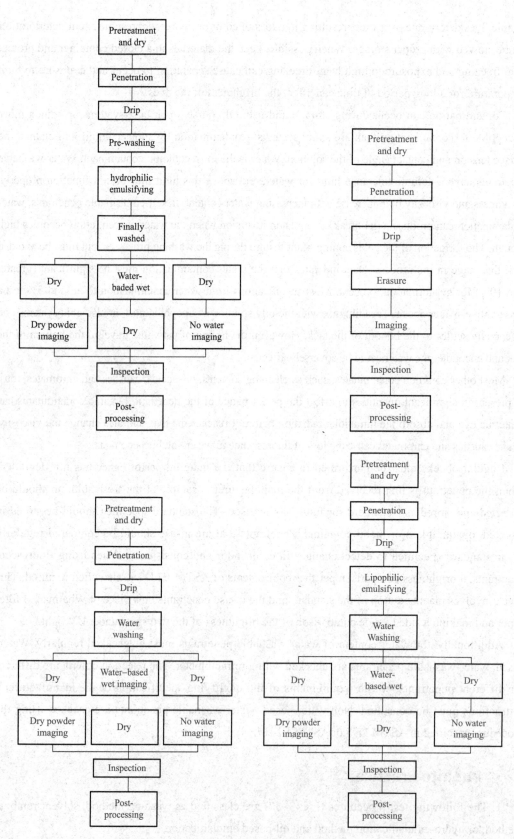

Fig. 3–37 Test Procedures

2. Look at the inspection flowchart (Fig. 3-38) to see what models of osmotics, removers and visible agents are? What is the penetration time and the visible time? What is the removal method?

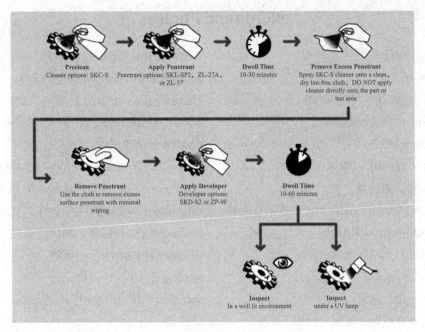

Fig. 3-38 The Inspection Flow Chart

3. Translation of the penetration detection report below(Table 3-5).

Table 3-5 PT Report of Welds

JOB NO.		JOB NAME		
PART NAME		DWG.NO.		
PART NO.		SIZE		
MATERIAL		QUANTITY		
SURFACE CONDITION		EXAMINATION STAGE		
EXAM.METHOD		EXAM. RATIO		
COMPARATOR BLOCK		CLEANING		
PENETRANT		DWELL TIME		
REMOVER		DRYING TIME		
DEVELOPER		DEV.TIME		
APPLICATION		EXAM.TEMP		
LIGHT (EQUIP.)		PROCEDURE (ID.& REV)		
SKETCH				
CODE		STANDARD		ACCEPT. LEVEL
CONCLUSION		REMARK		
OPERATOR/LEVEL/DATE		REVIEW/LEVEL/DATE		

Words and Phrases

quench [kwentʃ] vt. 将……淬火
diluted v. 稀释
dwell time 渗透时间；停留时间
stain [steɪn] n. 污点，瑕疵
ultraviolet lamp 紫外灯
toxicity [tɒk'sɪsəti] n. 毒性
testing kits 套装渗透液
fluorocarbon [ˌflʊroʊ'kɑːrbən] n. 碳氟化合物
hydrocarbon [ˌhaɪdrə'kɑːbn] n. 烃，碳氢
emulsification [ɪˌmʌlsɪˌfaɪ'keɪʃən] n. 乳化，乳化作用
halogens n.［化］卤素
liquid oxygen (LOX)（低温）液氧
oxidizing agent 氧化剂
a batch of 一批生产量，一批投料量
shot peening 喷丸硬化
buffing ［机］磨光，抛光
smearing 抹平，涂抹
fluid flow 液流
surface tension 表面张力
contact angle 接触角
surface wetting 表面润湿
capillarity [ˌkæpɪ'lærɪtɪ] n. 毛细管现象
capillary action 毛细作用
cohesive force 内聚力
hemispherical [ˌhemɪ'sferɪkl] adj. 半球的，半球状的
elastic force 弹力
elastic potential energy 弹性势能
thermodynamics [ˌθɜːməʊdaɪ'næmɪks] n. ［物］热力学
be analogous to 类似于……，与……相似
tangential [tæn'dʒenʃəl] adj. 切线的，正切的
normal ['nɔːml] 法线，垂直的

meniscus [mə'nɪskəs] n. 新月，半月，弯液面
acceleration of gravity 重力加速度
viscosity [vɪ'skɒsəti] n. 黏度，黏质，黏性
water-washable 水洗的
flow chart 流程图
lipophilic [ˌlɪpəʊ'fɪlɪk] adj. 亲油性的
hydrophilic [ˌhaɪdrəʊ'fɪlɪk] adj. 亲水性的
solvent-removable 溶剂去除的
precleaning n. 预洗漆，预清洗
anomaly [ə'nɒməli] n. 异物
welding flux 焊剂（料）
detergent [dɪ'tɜːdʒənt] n. 清洁剂，去垢剂，清洗剂
descaling n. 除鳞
abrasive cleaning 喷砂清理（法）
fluorescent penetrants 荧光渗透剂
visible or dye pentrants 着色渗透剂
lipophilic postemulsified 亲油性后乳化的
oil-based adj. 油基的
solvent-wipe 溶剂去除的
hydrophilic post-mulsifed 亲水性后乳化的
water-based adj. 水基的
dipping ['dɪpɪŋ] n. 浸涂
brushing n. 刷涂
spraying n. 喷涂
colorant ['kʌlərənt] n. 着色剂
soft-bristle brush 软毛刷
spray gun 喷枪
aerosol can 气溶胶罐
emulsifying agent 乳化剂
oil soluble 油溶性
water soluble 水溶性
solubility [ˌsɒljʊ'bɪləti] n. 溶解性
dispersant [dɪ'spɜːsənt] n. 分散剂

surfactant [sɜːˈfæktənt] n. & adj. 表面活性剂（的）
dissolve [dɪˈzɒlv] v. 溶解
redeposition n. 再沉淀
blotter [ˈblɒtə] n. 吸墨纸
dry powders developer 干粉显像剂
water soluble developer 水溶性显像剂
water suspendable developer 水悬浮型显像剂
solvent suspendable developer 溶剂悬浮型显像剂
fluffy [ˈflʌfi] adj. 松散的
electrostatic-powder 静电粉末
nonaqueous [nɔnˈeɪkwɪəs] adj. 非水的
volatile [ˈvɒlətaɪl] adj. 挥发性的

aqueous [ˈeɪkwɪəs] adj. 水的，水成的
permeate [ˈpɜːmɪˌeɪt] vt. 渗透，透过
color contrast 色对比度
incandescent [ˌɪnkænˈdesnt] adj. 白炽的
nonflammable [nɒnˈflæməbl] adj. 不燃烧的
odorless [ˈəʊdəlɪs] adj. 无味的
nontoxic [nɑnˈtɑksɪk] adj. 无毒的
leak testing 渗漏检测
resin [ˈrezɪn] n. 树脂
crater crack 坡口裂纹，火口裂纹
cutting tool 切削工具
carbide tool 硬质合金工具
halide [ˈhælaɪd] n. ［化］卤化物
filtered-particle penetrant 过滤性微粒渗透剂

A Sheet Work Manual

Dye Penetrant Inspection Procedure

This article provides you with a example dye penetrant inspection procedure. This procedure contains water washable process with both visible and fluorescent systems.

1. Scope

This procedure establishes the general requirements of penetrant testing of ferrous and non-ferrous components (essentially non-porous) and welds by water washable process, visible and fluorescent penetrant systems and are to be followed by NDT personnel third party.

2. Purpose

The requirements are intended to detect discontinuities that are open to test-surface and free from contaminants by interpreting the surface indications and evaluating them in accordance with the applicable referenced codes, standards or specifications by certified NDT personnel qualified in PT to Level Ⅱ.

Customer's approved specific Inspection/Evaluation manuals have also to be followed strictly, if available and applicable.

Any conflicts between any points of this procedure and the referenced documents or job specifications of customer shall be resolved (if given in writing) by the appointed ASNT NDT level Ⅲ or the divisional manager (NDT) or his authorised representative.

Where special circumstances require unique techniques of Penetrant Testing by Water-washable Process, specific technique sheets shall be attached as annex: Ⅳ etc., with the written approval of NDT Level Ⅲ.

3. References (Table 3-6)

Table 3-6 References

3.1	ANSI/ASME B31.1	Power Piping
3.2	ANSI/ASME B31.3	Chemical Plant and Petroleum Refinery Piping
3.3	ASME Sec. V	Boiler and Pressure Vessel Code
3.4	ASME Sec. VIII	Boiler and Pressure Vessel Code
3.5	ANSI/AWS DI.I	Structural Welding Code-Steel
3.6	API 1104	Standard for Welding Pipelines and related facilities
3.7	API 620	Design and Construction of Large, Welded Low Pressure Storage Tanks
3.8	API 650	Welded Steel Tanks for Oil Storage
3.9	ASTM E-165	Standard Test method for liquid penetrant test
3.10	ASTM E-1209	Standard Test Method for Fluorescent Penetrat Examination using Water-washable Process
3.11	ASTM E-1418	Test Method for Visible PT using Water-washable
3.12	ANSI/ASNT-CP-189	Personnel Qualification and Certification Standard
3.13	SNT-TC-1A of ASNT	Personnel Qualification and Certification Guidelines

4. Summary of test method & water washable penetrants

(1) A liquid penetrant which may be a visible or a fluorescent material is applied evenly over the surface being examined and allowed to enter open discontinuities. After a suitable dwell time, the excess surface penetrant is removed. A developer is applied to draw the entrapped penetrant out of the discontinuity and stain the developer. The test surface is then examined to determine the presence or absence of indications.

Note: Fluorescent penetrant examination shall not follow a visible penetrant examination because visible dyes may cause deterioration or quenching of fluorescent dyes.

(2) Processing parameters, such as surface precleaning, penetration time and excess penetrant removal methods, are determined by the specific materials used, the nature of the part under examination, (that is size, shape, surface condition, alloy) and type of discontinuities expected.

(3) Liquid penetrant examination methods indicate the presence, location and, to a limited extent, the nature and magnitude of the detected discontinuities. Each of the various methods has been designed for specific uses such as critical service items, volume of parts, portability or localised areas of examination. The method selected will depend accordingly on the service requirements.

(4) Water-washable Penetrants are designed to be directly water-washable from the surface of the test part, after a suitable penetrant dwell time. Because the emulsifier is "built-in" to the water-washable penetrant, it is extremely important to exercise proper process control in removal of excess surface penetrant to ensure against overwashing. Water-washable penetrants can be washed out of

discontinuities if the rinsing step is too long or too vigorous. Some penetrants are less resistant to overwashing than others.

(5) Post-emulsifiable penetrants with lipophilic and hydrophilic emulsifiers are not included in this procedure.

5. Equipment and materials

This procedure is intended for use with the following consumables or their equivalent. It has to be ensured that the consumables are selected in such a way that they are compatible to the test surface and object and as per test requirements.

There are two commonly used dye penetrant inspection systems, the specific information is shown in Table 3–7 and Table 3–8.

Table 3–7 Water-washable, Fluorescent Penetrant System

Manufacturer	Penetrant	Remover	Developer
Sherwin	HM-440, HM-430	Water	D-100, D-100NF
Ardrox	P133D, P134D, P135D	Water	9D1B
Magnaflux	ZL-60C	Water	SKD-LT or ZP-9F

Table 3–8 Water-washable, Visible Penetrant System

Manufacturer	Penetrant	Remover	Developer
Sherwin	DP-51	Water	D-100, D-100NF
Ardrox	906/303A	Water	9D1B
Magnaflux	SKL-WP	Water	SKD-NF, SKD-S, ZP-9B

Intermixing of penetrant materials from different families (manufacturers) is not permitted by this procedure. Manufacturer's recommendation for compatible penetrant systems must be adhered with.

Dye Penetrant Inspection—Control of Contaminants

When testing nickel based alloys, austenitic stainless steels and titanium, contaminant content of penetrant materials shall be controlled. Control shall be based on the manufacturers batch certification which shall include, as a minimum, the manufacturer's name, batch number and chemical contaminant content as determined in accordance with ASME Section V, paragraph T-625.

Only dye penetrant inspection materials having a batch number printed on the container and traceable to a valid manufacturer's batch certification on file shall be used.

6. Parts to be examined

This procedure shall be used for parts or welds in ferrous and non-ferrous materials in accordance with applicable code or specifications for parts or welds. When welds are tested, at least one inch of the base material on both sides of the weld is to be covered.

7. Surface preparation

Prior to the test, the area to be inspected and at least one inch on either side shall be free from all contaminants (dirt, grease, lint, slag, spatter, oil, scale, water and protective coatings) .

In general for welds, satisfactory results may be obtained when the surface to be inspected is in as welded condition. If mechanical methods of cleaning like grinding, machining or sanding is necessary, the surface area shall be etched to remove smeared metal (For full details, ASME Sec. V, article 24 and SE-165 shall be referred). After etching, suitable neutralising solutions shall be used and test surface be washed with water.

Unless otherwise recommended by the manufacturer, welded components or parts cleaned by vapour degreasing, organic solvents or detergents and properly protected from contamination, need not be re-cleaned with penetrant cleaner (remover) prior to application of the penetrant.

On occasion, a wire brush may be helpful in removing rust, surface scale, but it is used only when no other means of removal will surface. It shall be followed by cleaning with penetrant cleaner if compatible.

Drying: evaporation time, following precleaning, shall be 5 minutes minimum. Where indications of retained moisture exist, the evaporation time shall be increased till no evidence of moisture in the area of test can be detected.

8. Examination

(1) Temperature limitations:

The temperature of the penetrant materials and the surface of the part should be between 50 °F and 100 °F for fluorescent water washable penetrants and between 60 °F and 125 °F for visible penetrants.

(2) Penetrant application:

① Either immersion (dipping), flow-on, spray, or brushing technique is used to apply the penetrant to the precleaned dry specimen.

② The penetrant is applied evenly over the entire area.

③ Fillers shall be used on the upstream side of the air inlet when using compressed air to apply penetrant.

(3) Dwell times:

① Penetrant dwell times are critical and should be adjusted depending on temperature and other conditions and may require qualification by demonstration for specific applications.

② Typical minimum penetration times are given in the Table 3-9.

③ Penetrant shall remain on the test surface for the entire dwell time period.

④ Care shall be taken to prevent drying out of the applied penetrant and additional penetrant must be applied to re-wet the surface.

Table 3-9 Typical Minimum Penetration Times

Material	Form	Type of Discontinuity	Water-Washable Penetration Time*
Aluminium	Castings	Porosity, Cold Shuts	5 to 15 minutes
Aluminium	Extrusions, Forgings	Laps	NR**
Aluminium	Welds	Lack of Fusion, Porosity	30

continued

Material	Form	Type of Discontinuity	Water-Washable Penetration Time*
Aluminium	All	Cracks, Fatigue Cracks	30, not recommended for fatigue crack
Magnesium	Castings	Porosity, Cold Shuts	15
Magnesium	Extrusions, Forgings	Laps	not recommended
Magnesium	Welds	Lack of Fusion, Porosity	30
Magnesium	All	Cracks, Fatigue Cracks	30, not recommended for fatigue crack
Steel	Castings	Porosity, Cold Shuts	30
Steel	Extrusions, Forgings	Laps	not recommended
Steel	Welds	Lack of Fusion, Porosity	60
Steel	All	Cracks, Fatigue Cracks	30, not recommended for fatigue crack
Brass & Bronze	Castings	Porosity, Cold Shuts	10
Brass & Bronze	Extrusions, Forgings	Laps	not recommended
Brass & Bronze	Brazed Parts	Lack of Fusion, Porosity	15
Brass & Bronze	All	Cracks	30
Brass & Bronze			
Plastics	All	Cracks	5 to 30
Glass	All	Cracks	5 to 30
Carbide-tipped Tools	All	Lack of Fusion Porosity, Craks	30
Titanium & High Temp Alloys	All		not recommended
All Metals	All	Stress or Inter-granular Corrosion	not recommended

(4) Excess penetrant removal:

After the elapse of penetrant dwell time, the excess penetrant is removed by water spray. Water at 60 to 110 °F and a pressure not exceeding 30 psi (210 kPa) is applied with droplet type sprayer specifically designed for penetrant removal. The nozzle of sprayer is held so that water strikes the surface of the specimen at an angle of approximately 45 degrees. Care is to be taken to avoid over-washing, which causes washout of penetrant from discontinuities. Other methods of referenced codes or specifications could also be used if applicable for the test surface conditions.

(5) Drying:

The test surface must be dry prior to the application of non-aqueous or dry developers. If water-based wet developer is used, it is applied to still damp specimen immediately after the penetrant removal wash. Excessive heat or too long a drying time tends to bake the penetrant out of discontinuities.

(6) Developer application:

When the drying process is complete, the specimen is ready for the application of either dry or non-aqueous wet developer. When water-based wet developer is used, it is applied by flooding the surface to the wet specimen immediately after excess penetrant is removed.

It is recommended to use aerosol cans, after agitation, typically for weld inspection at various project-site works. Spray distance shall be 10 to 12 from test surface. The test-areas must be cool enough to prevent too repaid evaporation of the developer vehicle.

Dry developer is applied to the specimen by brushing with soft brush, by use of a powder gun, or by dipping the specimen in a tank of the developer and removing excess powder with a low pressure air flow. An even thin coat/film of developer is preferred. Applied developer shall not be removed from test surface.

(7) Dye penetrant inspection:

The area under inspection shall be observed during application of developer and at intervals during development time.

The recommended development time is between 7 and 30 minutes. Development time begins directly after application of dry developer and as soon as wet developer coating has dried on parts-surface.

Indications getting formed and formed at the test surface (by the blotting action of developer) be noticed, analysed (relevant indications be noted) under adequate lighting conditions.

Lighting conditions:

① Visible penetrant indications can be examined in either natural or artificial light. Adequate illumination is required to ensure no loss in the sensitivity of the examination. A minimum light intensity at the examination site of 100 fc (1 000 lx) is recommended.

② Examine fluorescent penetrant indications under black light in a darkened area. Visible ambient light should not exceed 2 ft candles (20 lx). The measurement should be made with a suitable photographic-type visible light meter on the surface being examined.

Black Light Level Control—Black light intensity, minimum of 1 000 μW/cm^2, should be measured on the surface being examined, with a suitable black light meter. The black light wavelength shall be in the range of 320 to 380 nm.

The intensity shall be checked at least once every 8 hours, or whenever the work station is changed. Cracked or broken ultraviolet (UV) filters should be replaced immediately. Defective bulbs, which radiate UV energy, must be replaced before further use. Since a drop in line voltage can cause decreased black light output with consequent inconsistent performance, a constant-voltage transformer should be used when there is evidence of voltage fluctuation.

Note:

① Certain high-intensity black light may emit unacceptable amounts of visible light, which will cause fluorescent indications to disappear. Care should be taken to use only bulbs certified by the supplier to be suitable for such examination purposes.

② The recommended minimum light intensity is intended for general usage. For critical examination, higher intensity levels may be required.

Black Light Warm Up—Allow the black light to warm up for a minimum of 10 min prior to its use or measurement of the intensity of the ultraviolet light emitted.

Visual Adaptation—The examiner should be in the darkened area for at least 5 minutes before examining parts to allow the eyes to adapt to the dark viewing.

③ Photochromic lenses shall not be worn during examination.

9. Interpretations and evaluations of indications

(1) All indications shall be evaluated in accordance of the referencing Code or Specification.

(2) Acceptance Criteria are listed for specific codes in Annex-1 of this document. Any conflicts between these documents and applicable Code section shall be resolved by the NDT Level Ⅲ or Divisional Manager (NDT).

(3) Mechanical discontinuities at the surface can result in false or irrelevant indications. Any indications which is believed to be non-relevant shall be regarded as a discontinuity and shall be re-examined to verify whether or not actual defects are present.

(4) Non-relevant indications and broad areas of pigmentation which would mask indications of defects are unacceptable and require corrective action by cleaning or other suitable means of surface preparation as described herein and retest.

(5) Make sure that surface indications are not false indications i.e. machining marks, mechanical conditions or other surface conditions which cause or produce false indications.

(6) Interpretation of indications found and determined to be rejectable shall be based on the size of the indication. Linear indications are those having a length greater than three times the width. Rounded indications are those that are circular or elliptical shape with the length equal to or less than three times the width.

(7) All examination shall be reported on the penetrant testing report form.

10. Safety

(1) The certified Inspector shall be responsible for compliance with applicable safety rules in the use of liquid penetrant materials.

(2) Liquid penetrant should not be heated or exposed to open flames.

(3) Penetrant materials may be highly volatile, relatively toxic and the liquid may cause skin irritation. Adequate ventilation at all times shall be used.

(4) Aerosol cans of penetrant materials should be kept out of direct sunlight or areas in excess of 130 °F. Excessive heat may cause aerosol cans to explode.

(5) Avoid looking directly into black light source, since the eyeball contains a fluid that fluoresces if black light shines directly into the eye.

11. Post cleaning

(1) Post cleaning is required (unless otherwise specified in contract) to remove any excess residues

from the penetrant process.

(2) A suitable cleaning technique such as water wash, vapours degreasing, solvent soak may be employed.

(3) Caution should be exercised to remove all developer prior to vapour degreasing as vapour degreasing can bake the developer on parts.

(4) Post cleaning of water washable fluorescent penetrant tests could be done by flushing surface with forced water spray or by flushing the surface with an approved solvent cleaner depending on the actual job requirements. Clean cloth and or absorbent paper towel and or dry air jet could also be used.

(5) Points mentioned in 11 (4) is applicable for water washable and visible dye also.

12. Personnel qualification and certification

(1) Personnel performing examinations in accordance with this procedure shall be qualified and certified in accordance with SNT-TC-IA (1992 edition) or CP-189 (1995 edition) of ASNT.

(2) Only individuals qualified to NDT Level I and working under the supervision of an NDT Level II or III or individuals qualified to NDT Level II shall perform the examinations in accordance with this procedure.

13. Preference for water–washable process

Water washable process shall be used for testing the following:

Articles having a rough surface.

Articles having threads and keyways.

High production of many small articles.

Weld inspection in closed vessels or tanks.

Weld or other component inspection when solvent process is found not suitable or preferable.

Articles having medium rough surfaces (if medium sensitivity is acceptable) .

Leak detection system.

Welds, articles when cost control is a prime point.

14. Characteristics of water-washable fluorescent penetrant tests

Advantages:

• Fluorescence ensures visibility

• Easily washed with water

• Good for volume testing of

• Small specimens

• Good on rough surfaces

• Good on keyways and threads

• Good on wide range of discontinuities

• Fast, single step process

• Relatively inexpensive

• Available in lox compatible form

Disadvantages:
- Requires darkened area for inspection
- Not reliable for detecting scratches
- Similar shallow surface discontinuities
- Not reliable on reruns of specimens
- Not reliable on anodised surfaces
- Acids and chromate's affect sensitivity
- Easily over-washed
- Penetrant subject to water contamination

Item 4 Visual Testing

Learning Objectives

1. Knowledge objectives
(1) To grasp the words, related terms and abbreviations about VT.

(2) To grasp the classification about VT system.

(3) To know the Instruments and Equipment of VT system.

(4) To know the testing procedure of VT system.

2. Competence objectives
(1) To be able to read and understand frequently used & complex sentence patterns, capitalized English materials and obtain key information quickly.

(2) To be able to communicate with English speakers about the topic freely.

(3) To be able to fill in the job cards in English.

3. Quality objectives
(1) To be able to self-study with the help of aviation dictionaries, the Internet and other resources.

(2) To do a good job of detection of safety protection.

4.1 Introduction of Visual Testing

Visual inspection (VT) (Fig. 4-1) is arguably the oldest and most widely used NDT method there is. For thousands of years, craftsmen have used their eyes to determine the quality of the products they made. In essence, this is still the case with visual inspection. This method involves the visual observation of the surface of a test object to evaluate the presence of surface discontinuities such as corrosion, misalignment of parts, physical damage and cracks. Visual testing can be done by looking at the test piece directly, or by using optical instruments such as magnifying glasses, mirrors,

Fig. 4-1 Visual Inspection (VT)

borescopes and computer-assisted viewing systems. VT can be applied to inspect castings, forgings, machined components and welds and is used in all branches of industry.

〖Point 1〗 Visual Testing (VT) Inspection Applications

Visual inspection is a useful tool to gauge the scope of corrosion, cracking, leaking and other damages, enabling to more effectively plan what inspection and maintenance services are required.

Visual inspection can be conducted on any assets that show visible forms of degradation, including:
- Piping
- Storage tanks
- Boilers
- Vessels
- Welds

A inspector will visually examine these assets and compare them to documented equipment drawings and previous inspection reports. The inspector makes note of any areas of concern, and depending on the findings, may recommend more advanced techniques to determine the damage's scope.

〖Point 2〗 Advantages and Limitations

VT advantages:
- Low cost
- Portable equipment (if any)
- Immediate results
- Minimum special skills required
- Minimum part preparation

VT limitations:
- Only suitable for surfaces than can be viewed
- Usually only larger defects can be detected
- Scratches can be misinterpreted for cracks
- Quality of inspection influenced by surface condition, physical conditions, environmental factors and physiological factors

〖Point 3〗 Visual Testing (VT) Inspection Standards

A standard on this list:
- American Petroleum Institute (API)
  - API 653 above Ground Storage Tank Inspection
  - API 510 Pressure Vessel Inspection
  - API 570 Piping Inspection

API 579 Fitness for Service

API 580 Risk Based Inspection

API 936 Refractory Inspection
- American Welding Society

Certified Weld Inspector (CWI)
- American National Standard Institute (ANSI)

ANSI N.45.2.6—Visual inspections for nuclear power plants
- ASME Boiler & Pressure Vessel Codes

SNT-TC-1A & CP189

〖Point 4〗 Definition of Visual Inspection

A non-destructive detection method for direct observation by the human eye or indirect observation and evaluation of items such as containers and metal structures and processing materials, parts and components, surface condition or cleanliness.

Visual inspection methods are usually divided into two categories: direct visual inspection and indirect visual inspection.

Direct visual inspection (Fig. 4–2): direct use of the human eye or the use of magnifying glass with a

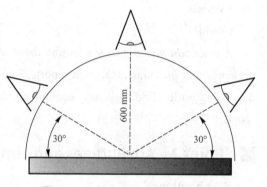

Fig. 4-2 Direct Visual Inspection

magnification of less than 6 times, the test piece for testing. So that the eye and the surface of the test piece is not more than 600 mm, the angle between the eye and the surface under examination is not less than 30 degrees.

Indirect visual inspection: the method of visual observation by optical instruments or equipment, which cannot be directly observed, is called indirect visual inspection; such as borescope (Fig. 4–3) detection.

Fig. 4-3 Borescope

〖Point 5〗 History and Development

Visual examination or testing (VT) is a method of non-destructive testing that has been neglected for much of its industrial application life. VT was the first non-destructive test (NDT) method used in the non-destructive testing industry, but was last method to be for mally acknowledged. Development of the visual method as an independent entity was fostered by the Electric Power Research Institute (EPRI) Nondestructive Examination (NDE) Center in the early 1980s. This was the result of the development of a training program for visual examination technology that included 120 hours of formal training. The need was prompted by the American Society of Mechanical Engineers, specifically Section XI-Rules for Inservice Inspection of Nuclear Power Plant Components. The program was designed to qualify personnel as visual examiners. Examination personnel scrutinizing the general condition of components were to comply with the requirements of the American Society for Non-destructive Testing, Recommended Practice No. SNT-TC-1A. Visual examination of components for general mechanical and structural conditions was satisfy to the requirements of ANSI N45.2.6. This standard was codified via the U.S. Federal Regulations, "Title 10, Code of Federal Regulations, Part 50, " requiring nuclear power plants to meet certain requirements for licensing. ASME sectored the visual examination into four categories based on the scope of inspection. The categories are classed as VT-1, VT-2, VT-3 and VT-4. VT-1 addresses the condition of a component, VT-2 the location of evidence of leakage, VT-3 the general mechanical and structural conditions of components and their supports, and VT-4 (which has been eliminated) focussed on the conditions relating to the operability of components or devices. Performance requirements for NDT are referenced in ASME Boiler and Pressure Vessel Code, Section V—Non-destructive Examination. Direct and remote visual testing is described in Article 9 of Section V—Visual Examination. Direct visual testing is definedas using "visual aids such as mirrors, telescopes, cameras, or other suitable instruments". Direct visual examination is conducted when access allows the eye to be within 25 inches (610 mm) of the surface to be examined, and at an angle not less than 30° to the surface to be examined.

Remote visual testing is divided into three categories: borescopes, fiberscopes and video technology. These have been developed chronologically. "Borescopes" also referred to as "endoscopes", were originally used to inspect the bores of rifles or cannons utilizing a hollow tube and mirror. The second generation of the endoscopes included are lay lens system in a rigid tube. This upgraded the image. Due to its rigid structure, endoscopes are limited to straight-line access, as is depicted in Fig. 4–4. Later innovations corrected this limitation by providing flexibility to the endoscopes. By 1955, the introduction of glass fiber bundles and fiber optic image transmission enabled the development of the fiberscope. Medical researchers experimented with different techniques in fiber optic image transmission during this period.

Imaging with fiber optic bundles decreased the clarity of the image transmission compared with the rigid lens systems of borescopes; however, this was a small price to pay for the opportunities it presented. The flexibility of the bundle opened up previously inaccessible areas to remote visual inspection, providing a more versatile tool to be used in industrial situations. This often eliminated the need to dismantle equipment for inspection. A typical fiberoptic borescope is illustrated in Fig. 4–5.

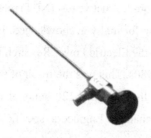 

Fig. 4–4　Rigid Borescope　　　　Fig. 4–5　Typical Fiberoptic Borescope

　　The evolution of the endoscope continued as the problems of eye fatigue associated with the use of endoscopes and fiberscopes prompted the development of various "add-on" cameras or closed circuit TV cameras that allowed for the display of images on a monitor. The first of such innovations was the tube-type camera. Many add-on camera systems are still presently in use, but due to their bulky exterior, smaller, solid-state imaging sensors, some of which are known as charge-coupled devices (CCDs), are replacing them. Fig. 4–6 provides a basic demonstration of how a charge coupled device (CCD) works. This new generation of CCDs stimulated a new wave of video endoscope technology. Small in diameter with high-resolution images, this new technology increased the range of industrial endoscopy applications. The physical size of the CCD as well as its ability to allow for electronic image processing and its other advantages broaden the application possibilities. One technological aspect worthy of mention is the CCD's ability to record images. Whether the camera is orthicon or vidicon tube technology or CCD technology, the present systems can record the images on videotape. With the advent of digital storage technology, the recording of images on other permanent media enhances the system's versatility.

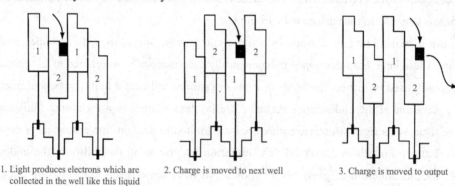

1. Light produces electrons which are collected in the well like this liquid　　2. Charge is moved to next well　　3. Charge is moved to output

Fig. 4–6　How Charge Coupling Works

Put into Practice

1. Translation.

　　Conduct automated visual inspections for defect detection in industrial applications (Fig. 4–7):

　　Visual inspection is the image-based inspection of parts where a camera scans the part under test for both failures and quality defects. Automated inspection and defect detection are critical for high-through put quality control in production systems. Visual inspection systems with high-resolution cameras efficiently detect microscale or even nanoscale defects that are difficult for human eyes to pick up. Hence, they are widely adopted in many industries for detection of flaws on manufactured surfaces such as metallic rails, semiconductor wafers and contact lenses.

Fig. 4-7　Visual Inspection for Defect Detection in Semiconductor Manufacturing

2. Read the following articles to answer questions about visual inspection objects, what the detection items and detection accuracy are, and how the inspection items are carried out.

Machine vision system for wings exception:

Test items: wings surface blemishes (surface oil, iron and other surfaces and product-independent color stains or scratches)

Reflexed wings, folded inside and outside fold, fold the flaps equivalent to a fixed-width, etc.

Detection accuracy: Accuracy $\geqslant$ 1 mm

Detection accuracy: Elastic waist missing, malposition, up and down moves >10 mm, left and right moves >10 mm; Malposition, upper and lower deviation > 10 mm, left and right deviation >10 mm, ends chipping >10 mm, monolithic material elastic waist width and standard contrast > $\pm$ 3 mm.

How Does the Detection System Work?

Machine vision system (Fig. 4-8) uses a CCD-based camera to capture a target image and converts it into image signal. Base on the image pixel distribution, brightness and its own color character in the image, it will be digitalized and characterized, such as stain, area, quantity, length ect., which are extracted according to a pre-defined routine by the image processing system. These features are then compared with stored reference patterns and output data, such as size, angle, quantity, qualified/disqualified, exist/non-exist, for achieving automatic recognition.

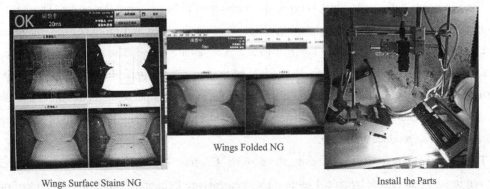

Wings Surface Stains NG　　　Wings Folded NG　　　Install the Parts

Fig. 4-8　Machine Vision System for Wings Exception

4.2 Theory and Principles

■ 〖Point 1〗 Characteristics of the Eye

In order to understand the physics of vision, it is necessary to first consider the characteristics of the eye. The eye can be compared to a radiation detector. Different wavelengths of light travel through the lens and reach the retina, which is located at the back of the eye. The rods and the cones of the retina in the human eye can sense wavelengths from about 400 nm up to approximately 760 nm. The eye performs the function of a spectrum analyzer that measures the wavelengths and intensity, as well as determining the origin of the light (from the sun or an artificial source). The light strikes the object to be viewed and is reflected towards the eye, through the lens and onto the retina as an image. The brain analyzes this image. The retina is similar to an array of tiny photosensitive cells. Each of these elements (cells) is connected to the brain through individual optic nerves. The optic nerves linking the eye to the brain can be compared to a bundle of electric cables. The major parts of the eye are shown in Fig. 4-9.

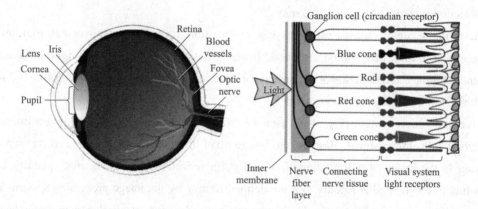

Fig. 4-9 Human Eye and Schematic View of the Retina

The iris opens and closes, thus varying the amount of light reaching the retina. The light then passes through the lens, which by changing shape, focuses the light and produces the image on the retina at the rear of the eye. Here a layer of rods and cones are found. The neurological connection from the rods and the cones pass through the rear of the eye via the optic nerve, which transmits the neurological signals to the brain. The brain processes the signals as perceptions of colors and details that vary in light intensity and color. It is necessary for a certain minimum level of light to be present before the eye can produce an image.

This level is known as the "intensity threshold". Contrast is something that shows differences between images placed side by side. Lighting requirements are frequently expressed in terms of ratios, due to the eye's ability to perceive a percentage of change rather than an absolute change in brightness.

The retina will only retain an image for a certain amount of time. This varies according to the size of the object and speed at which it is moving. The main limitations of vision are intensity threshold, contrast, visual angle, and time threshold. Visual acuity is the ability to distinguish very small details. For example, as the distance from the eye to the object increases, two lines that are close together appear as one heavy, dark line. The normal eye can distinguish a sharp image when the object being viewed subtends an arc of one-twelfth of a degree (five minutes), irrespective of distance from the eye to the object. Practically speaking, a person with "normal" vision would have to be within eight feet of a 20-inch TV monitor to resolve the smallest detail displayed. White light contains all colors. Newton proved that color is not a characteristic of an object but, rather, various wavelengths of light that are perceived as different colors by the eye. Color can be described as having three measurable properties: brightness, hue and saturation. The color of an object, ranging from light to dark, emitting more or less light, is known as brightness. Different wavelengths give us different perspectives of colors; this is known as hue. How green something is as opposed to white, is how saturated it is with green. In the United States NDT environment, visual acuity examinations are a requirement for certification. The visual inspector's natural visual acquity must be examined. The Jaeger (J) test is used in the United States for near-distance visual acuity. It consists of an English language text printed on an off-white card. The parameters for near-distance visual acuity are described in personnel certification and qualifications programs. Visual acuity requirements will vary depending upon the needs of specific industries.

〖Point 2〗 Sensitivity of the Human Eye

This page summarizes the basic characteristics of sensitivity of the human eye. This is useful when converting light intensities from luminous units like Lumens (lm) that take into account eye perception into radiometric units like Watts (W) that only take into account physical aspects.

A light source radiating 1 W of green light will appear much brighter than another source radiating the same amount of power of red light because the eye is more sensitive in the green region. The human eye is sensible to light wave which wavelength is roughly between 400 nm (violet) and 700 nm . Wavelengths shorter than 400 nm (Ultraviolet, UV) or longer than 700 nm (infrared, IR) are not visible.

The eye behaves differently in high or low light conditions: in daylight, for brightness levels above 3 cd/m^2 the vision is mainly done by the centre of the retina, we can see colors and the maximum sensitivity is at 555 nm (in the green region) . This type of vision is called photopic vision.

In low light conditions, for brightness levels below 30 $\mu cd/m^2$, the vision is mainly done by the peripheral region of the retina which is color-blind, while the centre region is not sensitive enough to see any color. This type of vision is called scotopic vision. Maximum sensitivity is at 507 nm (in the blue-green region) and red light is almost invisible.

The vision in-between photopic and scotopic vision is called mesopic.

The nice thing about scotopic vision and its reduced sensitivity to red light is that you can use a red flashlight to illuminate an object (to read a map, for example) without disturbing your nigh vision. It takes several minutes to get your vision used to the darkness; if you used a white flashlight instead, the eyes will switch back to photopic vision and you'll have to wait a few minutes again. Red light can directly be picked up by the centre of the retina without affecting night vision. For this reason, astronomers often use red flashlights.

The standard eye response is visible in Fig. 4–10 in both linear and logarithmic scale.

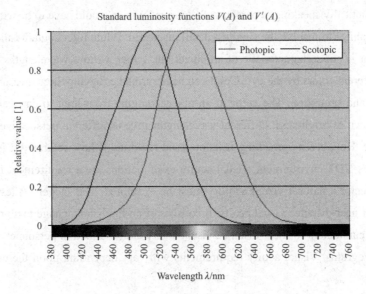

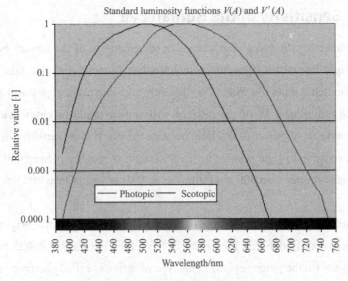

Fig. 4–10 Standard Eye Response

This standard eye sensitivity is also called standard luminosity function $V(\lambda)$ and is used, for photopic vision, to define a conversion between the radiated energy (in Watts) and the luminous flux (in Lumen). The standard luminosity function $V'(\lambda)$ refers to scotopic vision, but it shouldn't be used to convert to and from photometric units. A tabulated form of the two standard luminosity functions $V(\lambda)$ and $V'(\lambda)$ are represented in Table 4–1:

Table 4-1 Two Standard Luminosity Functions $V(\lambda)$ and $V'(\lambda)$

Wavelength λ /nm	Relative sensitivity		Equivalent of 1 W radiant power/(lm · W^{-1})	
	Photopic $V(\lambda)$	Scotopic $'(\lambda)$	Photopic $I_0 \cdot V(\lambda)$	Scotopic $I_0' \cdot V'(\lambda)$
390	0.000 1	0.002 2	0.068 3	3.74
400	0.000 4	0.009 3	0.273	15.8
460	0.060 0	0.567 2	41.0	964
500	0.323 0	0.981 7	221	1 669
507	0.449 0	1.000 0	307	1 700
510	0.503 0	0.996 6	344	1 694
520	0.710 0	0.935 2	485	1 590
550	0.995 0	0.480 8	680	817
555	1.000 0	0.404 8	683	688
560	0.995 0	0.328 8	680	559
600	0.631 0	0.033 2	431	56.4
650	0.107 0	0.000 7	73.1	1.19
680	0.017 0	0.000 1	11.7	0.170
690	0.008 2		5.60	
700	0.004 1		2.80	
750	0.000 1		0.068 3	
760	0.000 06		0.041 0	

For photopic vision, 1 W of radiant power at the wavelength of 555 nm is defined to correspond to a luminous flux of 683 lm. For scotopic vision, the sensitivity of the eye is greater and 1 W of radiant power at 507 nm corresponds to a luminous flux of 1 700 lm. These to wavelengths correspond to the maximum of each curve. Of course, every eye is different; these data correspond to the standard eye, as defined by the CIE 1931 (photopic) and CIE 1951 (scotopic) standards.

〖Point 3〗 Snell's Law

When light travels from one medium to another, it generally bends, or refracts. The law of refraction gives us a way of predicting the amount of bend. This law is more complicated than that for reflection, but an understanding of refraction will be necessary for our future discussion of lenses and their applications. The law of refraction is also known as Snell's Law, named for Willebrord Snell, who discovered the law in 1621.

Like with reflection, refraction also involves the angles that the incident ray and the refracted ray make with the normal to the surface at the point of refraction. Unlike reflection, refraction also depends on the media through which the light rays are travelling. This dependence is made explicit in Snell's Law via refractive indices, numbers which are constant for given media.

Snell's Law is given in Fig. 4-11.

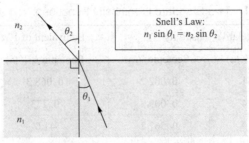

Fig. 4-11 Snell's Law

As in reflection, we measure the angles from the normal to the surface, at the point of contact. The constants n are the indices of refraction for the corresponding media.

Tables 4-2 of refractive indices for many substances have been compiled.

Table 4-2 Refraction for the Corresponding Media

n for Light of Wavelength 600 nm	
Substance	Refractive Index, n
Air (1 atmosphere pressure, 0 ℃)	1.000 29
Water (20 ℃)	1.33
Crown Glass	1.52
Flint Glass	1.66

Say, in our simple example above, that we shine a light of wavelength 600 nm from water into air, so that it makes a 30° angle with the normal of the boundary. Suppose we wish to find the angle x that the outgoing ray makes with the boundary. Then, Snell's Law gives:

$$1.33 \sin 30° = 1.000\ 29 \sin x$$

$$x = 41°$$

Refraction certainly explains why fishing with a rod is a sport, while fishing with a spear is not.

A more complicated illustration of Snell's Law (Fig. 4-12) proves something that seems intuitively correct, but is not obvious directly. If you stand behind a window made of uniform glass, then you know by now that the images of the things on the other side of the window have been refracted. Assuming that the air on both sides of your window have the same refractive indices, we have the following situation.

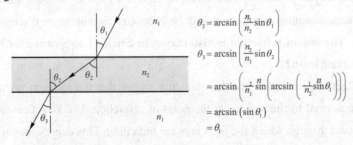

Fig. 4-12 More Complicated Illustration of Snell's Law

We find that the incoming and outgoing light beams are actually parallel.

Rearranging Snell's Law, with i and r being the incident and refracted angles.

$$n_1 \sin(i) = n_2 \sin(r)$$
$$(n_1/n_2) \sin(i) = \sin(r)$$

A qualitative description of refraction becomes clear. When we are travelling from an area of higher index to an area of lower index, the ratio n_1/n_2 is greater than one, so that the angle r will be greater than the angle i; i.e. the refracted ray is bent away from the normal. When light travels from an area of lower index to an area of higher index, the ratio is less than one, and the refracted ray is smaller than the incident one; hence the incident ray is bent toward the normal as it hits the boundary.

Of course, refraction can also occur in a non-rectangular object (indeed, the objects that we are interested in, lenses, are not rectangular at all) . The calculation of the normal direction is harder under these circumstances, but the behaviour is still predicted by Snell's Law.

Total Internal Reflection

An interesting case of refraction can occur when light travels from a medium of larger to smaller index. The light ray can actually bend so much that it never goes beyond the boundary between the two media (Fig. 4–13). This case of refraction is called total internal reflection.

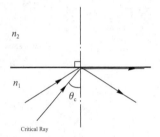

Fig. 4–13 Critical Angle

In the above diagram, imagine that we are trying to send a beam of light from a region with refractive index n_1 to a region with index n_2 and that $n_2 < n_1$. If x_1, x_2 are the angles made with the normal for the incident and refracted rays, then Snell's Law yields.

$$x_2 = \arcsin\left(\frac{n_1}{n_2} \sin x_1\right)$$

Since $n_2 < n_1$, we could potentially get an argument for the arcsin function that is greater than 1; an invalid value. The critical angle is the first angle for which the incident ray does not leave the first region, namely when the "refracted" angle is 90°. Any incident angle greater than the critical angle will consequently be reflected from the boundary instead of being refracted. For concreteness, pretend that we are shining light from water to air. To find the critical angle, we set $x_2 = 90°$. Using Snell's Law, we see that any incident angle greater than about 41° will not leave the water.

■ 〖Point 4〗 Inverse Square Law, General

Any point source which spreads its influence equally in all directions without a limit to its range will obey the inverse square law. This comes from strictly geometrical considerations. The intensity of the influence at any given radius r is the source strength divided by the area of the sphere. Being strictly geometric in its origin, the inverse square law applies to diverse phenomena. Point sources of gravitational force, electric field, light, sound or radiation obey the inverse square law. It is a subject of continuing debate with a source such as a skunk on top of a flag pole; will its smell drop off according to the inverse square law?

As one of the fields which obey the general inverse square law (Fig. 4-14), a point radiation source can be characterized by the relationship below whether you are talking about Roentgens, rads, or rems. All measures of exposure will drop off by inverse square law.

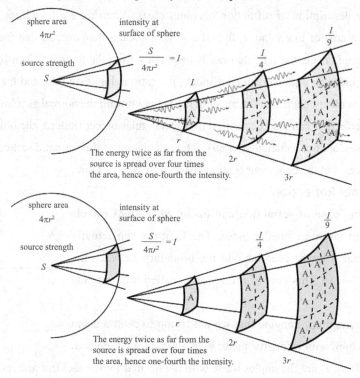

Fig. 4-14 Inverse Square Law

The source is described by a general "source strength" S because there are many ways to characterize a radiation source—by grams of a radioactive isotope, source strength in Curies, etc. For any such description of the source, if you have determined the amount of radiation per unit area reaching 1 meter, then it will be one fourth as much at 2 meters.

■ 〖Point 5〗 Others

Cleanliness

The amount of light that reaches the eye from an object is dependent on the cleanliness of the reflecting surface. In visual testing, the amount of light may be affected by distance, reflectance, brightness, contrast or the cleanliness, texture, size and shape of the test object. Cleanliness is a basic requirement for a successful visual test. Opaque dirt can maskor hide attributes, and excessively bright surfaces cause glare and prevent observation of the visual attributes.

Brightness

Excessive brightness within the field of view can cause an unpleasant sensation called glare. Glare interferes with the ability to see clearly and make critical observations andjudgments. Surface Condition Scale, rust, contaminants, and processes such as milling, grinding, and etching may affect the ability to examine a surface. This will be further discussed in the section relating to tools, equipment and accessories.

Shape

The shape of an object can influence the amount of light reflected to the eye, due to various angles that can determine the amount of light that will be reflected back to the eye.

Size

The size of an object will determine the type of scan pattern that may be used to view 100% of the object or it may determine that some magnification is necessary to get a closer view of details otherwise unobservable.

Temperature

Excessive temperature may cause distortion in viewing due to the heat wave effect. Most are familiar with the heat waves coming off a desert, resulting in a mirage; this is known as "heat wave distortion". In a nuclear reactor, underwater components are frequently distorted due to heat waves rising from the core that can interfere with the view from an underwater camera as it scans a particular component during visual examination.

Texture and Reflectance

One of the greatest variables in viewing an object is the amount of light that is reflected from it, and the angle at which the light strikes the eye. Excessive rust or roughness can cause diffusion of the light and limit the light returning to the eye. This can easily be corrected by increasing the amount of light or improving the surface condition of the object under examination.

Put into Practice

1. Read Fig. 4–15 to answer the wavelength range and composition of visible light, and what is the wavelength of the most sensitive light in the human eye.

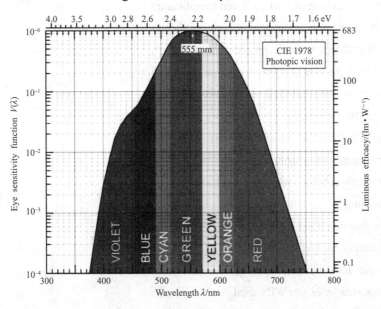

Fig. 4–15 Eye Sensitivity Function

VCD, left-hand ordinate and luminous efficacy, measured in lumens per watt of optical power (right-hand ordinate). VCD is maximum at 555 nm (after 1978 CIE data).

2. Translate Fig. 4-16 into Chinese.

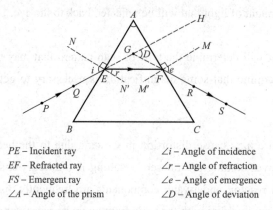

PE – Incident ray
EF – Refracted ray
FS – Emergent ray
$\angle A$ – Angle of the prism
$\angle i$ – Angle of incidence
$\angle r$ – Angle of refraction
$\angle e$ – Angle of emergence
$\angle D$ – Angle of deviation

Fig. 4-16 Incident Ray and Refracted Ray

4.3 Welding Terms

〖Point 1〗 VT of Welding

Visual Welding Inspectors Check List

1. Before welding commences
- Familiarization to the relevant code and specification
- Check welding equipment and calibration certificates
- Material identification, size, type and condition
- Consumables type, size, condition, storage and handling
- Review/witness WPS and PQR test and record
- Joint preparation (check)
- Welder qualification test (review/witness)

2. Welding process involved
- Check pre-heating before welding (if required)
- During welding
- Check weather condition
- Check clearance for welding/welder
- Check welder identification for weld
- Check consumables as per WPS used
- Check welding parameters as per WPS used

- Check distortion control
- Check interpass cleaning
- Check run out length (travel speed)
- Check interpass temperature
- Check usage of line up clamps
- Maintain daily log book

3. After welding completion
- Perform visual testing
- Weld and welder identification (check)
- Post weld heat treatment (if required)
- Non-destructive testing (witness)
- Acceptance standards of NDT
- Repairs (if any)
- Dimensional check (as per drawing)
- Document control—welding reports, etc.

〖Point 2〗 Summary of Duties

It is the duty of a Visual Welding Inspector to ensure all the welding and associated actions are carried out in accordance with the specification and any applicable procedures.

A Welding Inspector must:

Observe
To observe all relevant actions related to weld quality throughout production.

Record
To record, or log all production testing points relevant to quality, including a final report showing all identified imperfections.

Compare
To compare all recorded information with the acceptance criteria and any other relevant clauses in the applied application standard.

〖Point 3〗 Visual Welding Inspectors Equipment

Measuring device, e.g.
- Flexible tape, steel rule
- Temperature indicating crayons
- Welding gauges, e.g. TWI multi-purpose gauge
- Voltmeter
- Ammeter
- Magnifying glass
- Torch/flash light

〖Point 4〗 Joint Terminology (Fig. 4-17~Fig. 4-19)

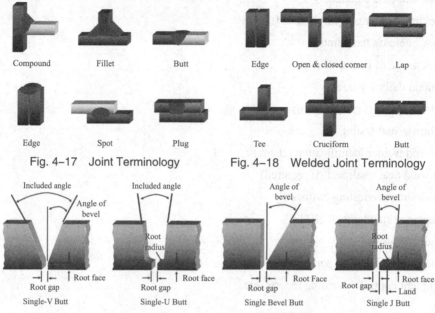

Fig. 4-17　Joint Terminology

Fig. 4-18　Welded Joint Terminology

Fig. 4-19　Types of Joint Preparation

Single sided preparations (Fig. 4-20) are normally made on thinner materials, or when access form both sides is restricted. Double sided preparations (Fig. 4-21) are normally made on thicker materials, or when access form both sides is unrestricted.

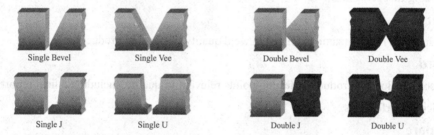

Fig. 4-20　Single Sided Joint Preparation　　Fig. 4-21　Double Sided Joint Preparation

The weld toe (Fig. 4-22) is located at the junction of the weld surface and the base mental, and the wrong angle (Fig. 4-23) of the weld toe will cause stress concentration.

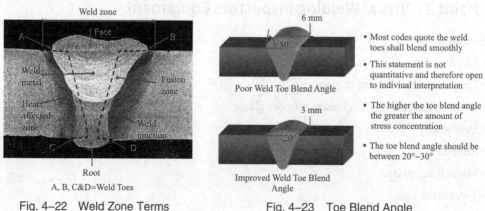

Fig. 4-22　Weld Zone Terms　　　　　Fig. 4-23　Toe Blend Angle

〖Point 5〗 Weld Defects (Fig. 4-24)

Defects which may be detected by visual testing can be grouped under five headings:
Root Defects, Contour Defects (Fig. 4-25), Surface irregularities, Surface cracks(Table 4-3), Miscellaneous.

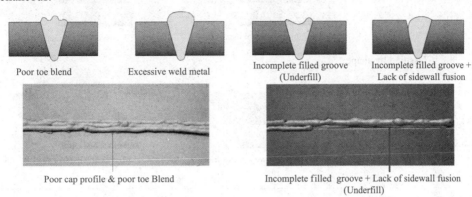

Fig. 4-24 Weld Defects

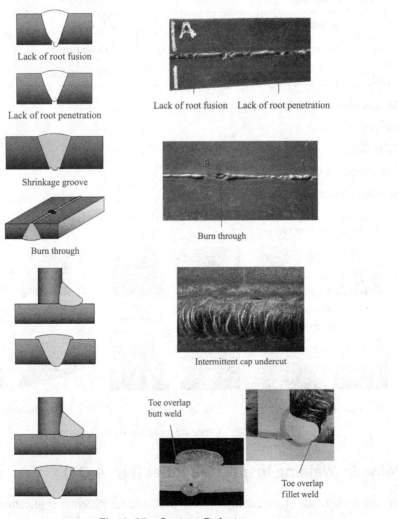

Fig. 4-25 Contour Defects

133

Table 4–3 Classification of crack

Cracks	Classified by shape	Classified by position	
		Longitudinal parent metal crack	Transverse weld metal crack
		Longitudinal weld metal crack	Lamellar tearing
	Longitudinal	HAZ	
	Transverse	Centerline	
	Branched	Crater	
	Chevron	Fusion zone	
		Parent metal	

Other associated weld defects (Fig. 4–26):

Cavities

Solid inclusions

Set up irregularities

Parent material defects

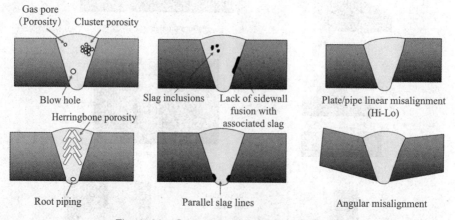

Fig. 4–26 Other Associated Weld Defects

▪ 〚Point 6〛 Welding Inspection Ruler (Fig. 4–27)

There are a multitude of measuring devices available for different applications. For the sake of brevity, only a few will be discussed here. The direct visual inspection method is frequently augmented by the use of several common tools used to measure dimensions, discontinuities, or range of inspection. Among these are linear measuring devices, outside diameter micrometers, ID/OD calipers, depth indicators, optical comparators, gauges, templates, and miscellaneous measuring devices.

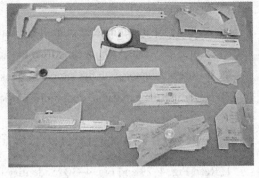

Fig. 4-27 Inspection Ruler

(1) The uses, measurement range and technical parameters of welding calipers are as shown in Fig. 4-28.

The product mainly consists of a main scale, a slider and a multi-purpose gauge. It is a weld detention gage used to detect the bevel angle of weldments, the height of various weld lines, weldment gaps and the plate thickness of weldments.

It is suitable for manufacturing boilers, bridges, chemical machinery, and ships and for inspecting the welding quality of pressure vessels.

This product is made of stainless steel, with reasonable structure and beautiful appearance, which is easy to use.

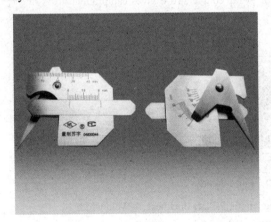

Measuring Items		Range	Tolerance for Indicating Value
Welding Thickness		0-40	±0.2
Weld Line Height	Multi-purpose Gauge	0-20	±0.3
	Slider	0-20	±0.2
Weld Gap		1-3	±0.2
Weldment Bevel Angle		0°-90°	30'

Fig. 4-28 Welding Calipers

(2) Instructions for use:

The edge scale can be used as straight steel ruler to detect the thickness of 0-40 mm sheets.

The multi-purposed gauge is used to measure the height of butt weld lines. The indicator on the multi-purpose gauge corresponding to the scale on the main scale is the height of the butt weld line.

The slider is used to measure the height of fillet welds. The indicator on the slider corresponding to the scale on the main scale is the height of the fillet weld.

In measuring the height of 45-degree-angle weld lines, the indicator on the slider corresponding to the scale on the main scale is the height of the 45-degree-angle weld line.

In measuring the gap of weldments, the indicator on the multi-purpose gauge corresponding to the scale on the main scale is the gap of the weldment.

(3) Maintenance:

① Welding inspection ruler cannot be stacked together with other tools, to avoid distortion, scratches and fuzzy scale, which can affect the accuracy.

② It is forbidden to scrub the engraved lines with banana water.

③ Never use the gap gauge on a multi-purpose ruler as a screwdriver.

Put into Practice

1. Read Fig. 4–29 and Fig. 4–30 to identify the butt welding features.

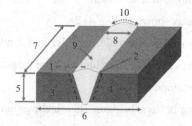

Fig. 4–29 Butt Weld Features

1-Cap height; 2-Weld toes; 3-Fusion junction; 4-HAZ; 5-Plate thickness; 6-Plate width;
7-Plate & weld length; 8-Weld width; 9-Cap contour; 10-Convex cap profile

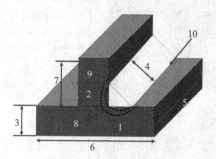

Fig. 4–30 Fillet Weld

1-Horizontal leg length; 2-Vertical leg length; 3-Plate thickness; 4-Weld width;
5-Plate/weld length; 6-Plate width; 7-Plate height; 8-Root; 9-HAZ; 10-Weld toe

2. Read Fig. 4–31 to identify the welding defects.

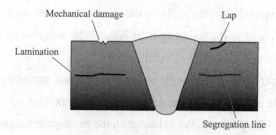

Fig. 4–31 Defects

4.4 Industrial Endoscope

Whenever the eye cannot obtain a direct, unobstructed view of the specimen test surface without use of another tool, instrument, or device, a "remote visual" examination is performed. Recall that a direct visual examination is an examination that can usually be made when access is sufficient to place the eye within 24 inches (610 mm) of the surface to be examined and at an angle not less than 30° to the surface to be examined. Most codes permit the use of mirrors and magnifying lens to assist with direct visual examinations. A remote visual examination can be defined as an examination that uses visual aids such as mirrors, telescopes, borescopes, fiber optics, cameras, or other suitable instruments.

An endoscope (sometimes called a borescope, videoscope, inspection borescope camera) is a type of precision instrument composed of a display monitor and a flexible insertion tube with an optical lens and micro-camera sensor. Its function is to have a remote visual inspection of those narrow areas that are inaccessible to human eyes, which is widely applied in aviation, ship, chemical, engine, power, boiler, automotive industry, etc (Fig. 4–32).

   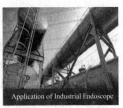

Fig. 4–32 Aplication of Industrial Endoscope

■ 〖Point 1〗 Endoscope Components

Traditional endoscopes comprise an airtight and waterproof elongated tube having a distal end with an objective lens for imaging and a proximal end with an eyepiece for viewing. The elongated tube includes a relay lens system, or a fiber bundle, to transmit the image formed by the objective lens to the proximal end of the tube. The function of the eyepiece is to magnify the image at the proximal end for the observer. Endoscopes used in clinical applications further consist of a work channel and an irrigation channel.

In the last five decades, the introduction of optical fibers, rod lenses, and then the electronic detector has advanced the development of endoscopes dramatically. Various endoscopes have been developed for general applications and to meet some special requirements. Some commonly used endoscopes are arthroscopes to examine joints, bronchoscopes to examine air passages and lungs, colonoscopes to examine the colon, cystoscopes to examine the urinary bladder, gastroscopes to examine the small intestine, stomach and esophagus, hysteroscopes to examine the uterus.

In some form or another, all endoscopes use optical elements to guide light to a target and transfer an image to the eye or to a detector. The basic optical elements of an endoscope include an illumination system, an imaging system, an image transmission system (relay system), and a viewing system (eyepiece or electronic sensor). Fig. 4–33 is a diagram of a conventional optical system in a rigid endoscope where, for simplicity,

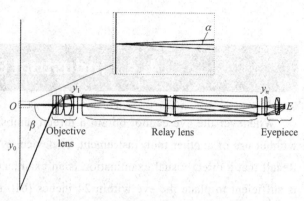

Fig. 4–33 Endoscope Components

only one relay stage is shown. The system consists of three basic and separate optical components, as follows:

Typical optical layout of a conventional endoscope with a one-stage relay lens. It consists of three lenses: the objective lens, relay lens, and eyepiece. Here y_0 is the height of the field, y_1 is the first intermediate image height, and y_n is the final image height through the relay lens. E is the exit pupil of the eyepiece, and it is the location for placement of the observer's pupil.

An objective lens, which forms the first inverted intermediate image y_1 of the object y_0. A relay lens system, which reimages the first intermediate image y_1 to the final image y_n at the proximal end of the endoscopic tube. In most endoscopes, there are several unit-magnificationrelay lenses forming the intermediate images y_2, y_3, …, and the final image y_n. An eyepiece or a focusing lens, which presents the final image y_n to a sensor. Traditionally, an eyepiece is attached to the endoscope and produces a magnified virtual image for the observer. In modern endoscopes, a camera lens is attached to the endoscope, and it produces a real image on the electronic sensor. The image is then displayed on a monitor or other display device. According to optical transmission systems, we can classify endoscopes into three groups: rigid, fiber optic and video. In recent years, some new types of endoscopes, such as wireless, scanning, and stereo endoscopes, have been developed.

Figure illustrates a basic optical configuration at the distal end of an endoscope. The light from the fiber bundle illuminates the observation region. Part of the reflected light from the tissue is captured by the objective lens. There are two major obstacles to designing an endoscope having a small diameter. The first obstacle is the lack of sufficient illumination. An endoscope with a small diameter does not provide enough space to transmit light. The second obstacle is light collection efficiency. The aperture is small because of the small diameter of the endoscope, which results in a low light collection efficiency. The amount of light captured by the eye or an electronic sensor is principally determined by three factors: the intensity of the light incident upon the observation region; the optical characteristics of the observation surface, such as surface reflectivity and curvature; the light collection efficiency of the objective lens and the transmission of the optical system.

Basic optical configuration at the distal end of an endoscope. The light from the fiber bundle illuminates the tissue. Part of the reflected light is captured by the objective lens.

〖Point 2〗 Classification

Rigid Endoscopes

As the name suggests, rigid endoscopes (Fig. 4–34) have a rigid tube to house the refractive relay lenses and illumination fibers. The relay lenses transfer the image at the distal end to the proximal end of the tube so that the image can be directly viewed through an eyepiece or with an imaging detector. When the endoscope is designed for direct view through the eyepiece, an odd number of relay stages is required in order to correct for the inversion produced by the object because the eyepiece has a positive magnification and does not produce an inverted virtual image. When an electronic sensor is used in an endoscope, the parity of the number of relay stages is not important since the inversion can always be performed electronically or through software.

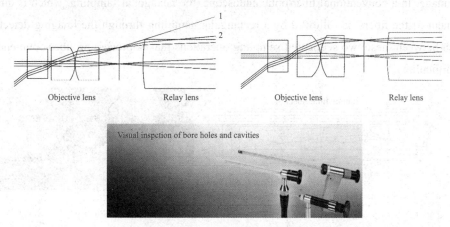

Fig. 4–34 Rigid Endoscopes

Flexible Endoscopes

Flexible endoscopes use imaging fiber bundles to transfer the image from the distal end of the endoscope to the imaging lens or eyepiece, as shown in Fig. 4–35. With fiber bundles, the diameter of the space required for the image relay is reduced significantly, leaving more space for other instrument channels. Other advantages of fiber-optic endoscopes include transmitting the image over long distances and observing around corners.

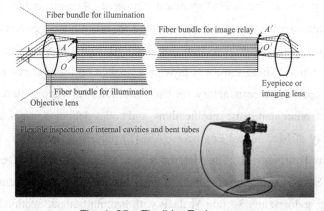

Fig. 4–35 Flexible Endoscopes

Video Endoscopes

One of the most important advances in endoscopes is the introduction of video chip technology on the distal end of the device. Image relay is not necessary given that the objective lens images the observation surface directly to the sensing elements in the imaging detector. As shown in Fig. 4–36, the imaging head of a video endoscope generally consists of an imaging detector, objective lens, and illumination optics. Two types of imaging detectors, CCD and CMOS, are commonly used in video endoscopes. Video endoscopes have revolutionized many surgical procedures by improving visualization while minimizing the tissue damage that is generally caused by invasive approaches.

The advantages of a video endoscope over a fiber-optic endoscope include the number of pixels in an imaging detector is typically greater than the number of fibers in a fiber bundle, giving a sharper image. In a conventional fiber-optic endoscope, the hexagonal sampling, which is due to the arrangement of the fibers, is followed by a rectangular sampling through the imaging detector. The mismatch between these two sequential sampling operations makes processing fiber-optic endoscope images difficult.

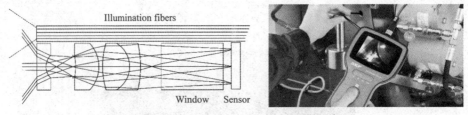

Fig. 4–36　Video Endoscopes

■ 〖Point 3〗 Standards, Codes and Specifications

The above applications all require VT and RVT to detect surface anomalies. A common source for material specifications is the American Society for Testing and Material (ASTM) standards. ASTM was founded in 1898 and is a scientific and technical organization formed for "the development of standards on characteristics and performance ofmaterials, products, systems and services; and the promotion of related knowledge." At this time, the annual ASTM Standards fill 75 volumes and are divided into 16 sections. The ASTM standard represents a common viewpoint of producers, users, consumers, andgeneral interest groups intended to aid industry, government agencies, and the generalpublic. Two metal material sections are Section1 Iron and Steel Products and Section 2 Nonferrous Metal Products. These standards provide guidance on the material conditions that must exist in order to be considered satisfactory for use. Additionally, when material is fabricated and formed into a product, other standards, specifications, and codes delineate visual testing requirements. The American Society of Mechanical Engineers (ASME) publishes the ASME Boiler and Pressure Vessel Code, Sections Ⅰ through Ⅺ. The Material section is Section Ⅱ. The Design sections are Ⅰ, Ⅲ, Ⅳ, Ⅷ, and Ⅹ. Section Ⅴ contains the methodology for non-destructive examination and Section Ⅸ addresses welding and brazing. Section Ⅵ deals with heating boilers, Section Ⅷ, pressure vessels and Section Ⅺ, in-service inspection of nuclear power plant components. Scattered throughout these sections are visual examination requirements. The American National Standards Institute (ANSI) is a consensus-approval-based organization. ANSI's B31.1 Power Piping and B31.7 Nuclear Power Piping also provide visual examination requirements for materials, fabrication and erection. The American Petroleum Institute (API) has developed approximately 500 equipmentand operating standards relating to the petroleum industry that are used worldwide. An example is the API standard for Welding of Pipelines and Related Facilities (API 1104). The aerospace and military standards are being replaced with the more commonly accepted industry codes and standards.

■ 〔Point 4〕 Testing Methods

Visual examinations and other non-destructive test methods cover the spectrum of examining materials from raw product form to the end of their useful lives. Initially, when raw material is produced, a visual examination is conducted to locate inherent discontinuities. As the material is further transformed through the manufacturing process, a product results. At this stage, the visual examination method is used to find discontinuities that are produced during the primary processing steps. When the product is further developed into its final shape and appearance, the secondary processes that give the product its final form can also introduce new discontinuities. Finally, the product is placed into service and is subject to stresses, corrosion, and erosion while performing its intended function. The process concludes when the material has reached the end of its useful life and is removed from the source. At every stage, the visual examination method is applied using various techniques to ascertain the physical condition of the material that became the component, system or structure serving the needs for which it was intended. After material is produced, visual examination is used to assure that a product will meet the specification requirements prior to processing into a product form for use in its intended service. The technology associated with visual testing (VT) and remote visual testing (RVT) includes a spectrum of applications, including various products and industries such as:

- Tanks and vessels Buildings
- Fossil-fuel power plants
- Nuclear power plants Turbines and generators
- Refinery plants
- Aerospace

Tanks and vessels usually contain fluids, gases or steam. Fluids may be as corrosive asacid or as passive as water, either of which can cause corrosion. Tank contents are not always stored at high pressure. Conversely, vessels usually contain substances under substantial pressure. This pressure, coupled with the corrosive effects of fluids and thermal or mechanical stresses, may result in cracking, distortion, or stress corrosion of the vessel material. Buildings also serve as a source for a myriad of RVT applications. These applications include location of clogged piping; examination of heating and cooling (HVAC) heat exchangers; and looking for cracking, pitting, blockages and mechanical damage to the components. Structural damage that may be present in the support systems, beams, flooring or shells, such as cracking, corrosion, erosion or warpage can also bedetected. Fossil-fuel power plants have piping, tubing, tanks, vessels, and structures that are exposed to corrosive and erosive environments as well as to other stresses. These components may require RVT. Turbines and generators, existing at both fossil-fuel and nuclear power plants are vulnerable to damage due to high temperatures, pressures, wear, vibration, and impingement of steam, water, or particles. Accessing the small openings and crevices to reach damaged turbine blades becomes a very tedious job and a serious challenge, but the effort of performing remote inspections through limited access ports reduces the need and cost of downtime and disassembly of major components. VT and RVT technologies and techniques are used in nuclear power plants as well.Water used for shielding and cooling is exposed to both ionizing radiation and radioactive surface contamination. The use of water as a coolant and radiation shield in a nuclear environment places additional requirements on RVT evaluation. The equipment must not only be waterproof, but also tolerant of radioactive environments. Due to process requirements in refineries, the containment of pressure and temperature is a necessity of paramount importance, as is the containment of hazardous materials. These same materials can be a source of corrosion to piping, tanks, vessels, and structures, all of which are in constant need of monitoring.

The VT technique most often used to detect inherent discontinuities is direct visual. The human eye assisted by measuring devices, auxiliary light sources, visual aids (e.g. magrs and mirrors), and the recording media of photographs and sketches are most commonly used for this application.

Observe the flaw in Fig. 4–37.

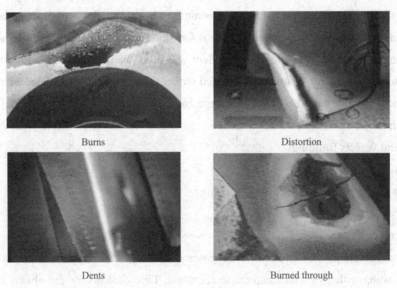

Fig. 4–37 Flaw

Generally, the main limitation to visual testing is access. The image of the object must be delivered to the eye. That image is always of the surface of an object. Visual testing is capable of examining the surface of an object unless the material is translucent. Remote visual testing advances are being driven today, as in recent years, by consumer demand and improvements in video technology. The challenge remains to understand fully "what" the inspector is examining and "how" the image is delivered to the eye. As designers make the image-gathering package smaller and smaller, the limitations of access will be further reduced. Applications in the field of medicine have been influencing the industrial field for years. Military applications including drones and robotic devices should continue to bring innovations to the technology of remote visual testing.

〖Point 5〗 Essential Steps for Flexible Endoscope Reprocessing

To ensure flexible endoscopes are safe for patient use, all staff involved in reprocessing this equipment must understand and consistently follow a number of steps which have been distilled down to seven essential steps. Ensuring adherence to these steps requires a complete and effective reprocessing program. These recommendations apply to all settings where endoscopic procedures are performed and where endoscopes are reprocessed.

1. Pre-cleaning

Pre-clean flexible endoscopes and reusable accessories by following the device manufacturer's instructions for use (IFU).

Perform pre-cleaning immediately following completion of the endoscope procedure to help prevent the formation of biofilm.

2. Leak-testing

For endoscopes that require leak testing, perform the leak test using manufacturer's IFU after each use and prior to manual cleaning. Leak testing detects damage to the external surfaces and internal channels of the endoscope that can lead to inadequate disinfection and further damage of the endoscope.

3. Manual cleaning

Perform meticulous manual cleaning including brushing and flushing channels and ports consistent with the manufacturer's IFU before performing high-level disinfection (HLD) or sterilization. Perform manual cleaning within the timeframe specified in the manufacturer's IFU. Manual cleaning is the most critical step in the disinfection process since residual organic material can reduce the effectiveness of HLD and sterilization.

4. Visual inspection

After manual cleaning, visually inspect the endoscope and its accessories. Visual inspection provides additional assurance that the endoscope and its accessories are clean and free of defects. Complex devices such as flexible endoscopes may require the use of lighted magnification or additional methods to assist with the inspection process.

5. Disinfection or sterilization

Following cleaning and visual inspection perform HLD or sterilization in accordance with the manufacturer's IFU. Carefully review and adhere to the endoscope manufacturer's reprocessing instructions and to the IFU for chemicals or sterilants and any equipment (e.g. automated endoscope reprocessors) used for reprocessing to help ensure that effective disinfection occurs.

6. Storage

After reprocessing is complete, store endoscopes and accessories in a manner that prevents recontamination, protects the equipment from damage, and promotes drying. Store processed flexible endoscopes in a cabinet that is either:

① Of sufficient height, width and depth to allow flexible endoscopes to hang vertically without coiling and without touching the bottom of the cabinet;

② Designed and intended by the manufacturer for horizontal storage of flexible endoscopes.

7. Documentation

Maintain documentation of adherence to these essential steps each time an endoscope is reprocessed. Documentation is essential for quality assurance purposes and for patient tracing in the event a look back is necessary.

Put into Practice

1. Translate Table 4-4 of parameters for the endscope into Chinese.

Table 4-4 Endoscope Performance Parameters

Designation	TKES 10F
Display unit	Included
Insertion tube & light source	Flexible tube
Image sensor	CMOS image sensor
Resolution ($H \times V$) —Still image (static) —Video (dynamic)	640 pixels×480 pixels 320 pixels×240 pixels
Size tip (insertion tube) diameter	5.8 mm (0.23 in.)
Tube length	1 m (39.4 in.)
Field of view	67°
Depth of field	1.5-6 cm (0.6-24.4 in.)
Light source	4 white adjustable LED (0-275 lx/4 cm)
Probe working temperature	-20 to +60 ℃ (-4 to +140 °F)
Ingress protection level	IP 67
Case dimensions	360 mm×110 mm×260 mm (14.2 in.×4.3 in.×10.2 in.)

2. Read the following essay to answer questions about the type of defect detected by the pipeline visual inspection, the speed of detection, and the frequency of detection.

How Can You Use Eddy Current NDT for Tube Inspection?

Tubes may be inspected using eddy current (ECT) non-destructive testing (NDT) from the outer diameter (OD), usually at the time of manufacture and from the inner diameter (ID), usually for in-service inspection, particularly for heat exchanger inspection (Fig. 4-38).

Fig. 4-38 ID Heat Exchanger Tube Testing

Heat exchangers used for petrochemical or power generation applications may have many thousands of tubes, each up to 20 m long. Using a differential Internal Diameter (ID or "bobbin") probe, these tubes can be tested at high speed (up to 1 m/s with computerised data analysis) and by using phase analysis, defects such as pitting can be assessed to an accuracy of about 5% of tube wall thickness. This allows accurate estimation of the remaining life of the tube, allowing operators to decide on appropriate action such as tube plugging, tube replacement or replacement of the complete heat exchanger.

The operating frequency is determined by the tube material and wall thickness, ranging from a few kHz for thick-walled copper tube, up to around 600 kHz for thin-walled titanium. Tubes up to around 50 mm diameter are commonly inspected with this technique. Inspection of ferrous or magnetic stainless steel tubes is not possible using standard eddy current inspection equipment.

Dual or multiple frequency inspections are commonly used for tubing inspection, in particular for suppression of unwanted responses due to tube support plates. By subtracting the result of a lower frequency test (which gives a proportionately greater response from the support) a mixed signal is produced showing little or no support plate indication, thus allowing the assessment of small defects in this area. Further frequencies may be mixed to reduce noise from the internal surface.

3. Translate the following process cards in Table 4-5.

Table 4-5   ET Report

Product name		Product No.	
Weld types		Welding method	
Material		Type of illumination	
Inspection tools	Weld inspection ruler, luxmeter		
Items	Standard requirement	Inspection results	
Surface crack	Not allowed		
Surface porosity	Not allowed		
On the surface of slag	Not allowed		

		continued
Leakage welding	Not allowed	
Burn through	Not allowed	
Overlap burr	Not allowed	
Spatter	Not allowed	
Inspector:		Date:

Words and Phrases

Homogeneity n. 同种，同质，均匀性，一致性
discontinuity [ˌdɪsˌkɒntɪˈnjuːəti] n. 不连续性，间断，不均匀性
void [vɔɪd] n. 孔洞
lap [læp] v. 搭接
undercut [ˌʌndəˈkʌt] n. 焊接咬边，下凹
anisotropy n. 各向异性，非匀质性
forging [ˈfɔːdʒɪŋ] n. 锻造
gas void or gas porosity 气孔
cold shut 冷隔
scab [skæb] n. 痕，瑕，疵
lack of fusion 未熔合
fatigue crack 疲劳裂纹
lack of penetration 未焊透
remelt [riːˈmelt] v. 再融化，再熔化
consolidation n. 巩固，合并
shrinkage porosity 缩松，松心
microporosity [ˌmaɪkrəʊpɔːˈrɒsəti] n. 微孔性，微孔率，[冶] 显微缩松，显微疏松（指非常微小，只有借助显微镜才可见到的孔隙）
shrinkage crack 缩裂
wormholes n. 虫孔，蜗孔
blistering [ˈblɪstərɪŋ] n. 气孔，针孔
residual stress 残余应力
autogenous welding 气（乙炔）焊
metal arc 金属电弧
submerged arc 埋弧

arc welding 电弧焊
heat-affected zone (HAZ) 热影响区
underbead crack 焊根裂缝，焊道下裂缝
martensite [ˈmɑːtɪnˌzaɪt] n. [冶] 马氏体
ferritic steel 铁素体钢
high-carbon steels 高碳钢
medium-carbon steel 中碳钢
stainless steel 不锈钢
quenching stress 淬火应力
mitigation [ˌmɪtɪˈgeɪʃn] n. 缓解，减轻，平静
stress corrosion cracking 应力腐蚀裂纹[开裂]
precipitate [prɪˈsɪpɪteɪt] n. 沉淀物
cleavage [ˈkliːvɪdʒ] n. 劈开，分裂
hydrogen cracking 加氢裂化，氢压下裂化
weld toe 焊趾
convexity [kɒnˈveksəti] n. 凸状，凸面
concave adj. 凹的，凹入的 n. 凹，凹面
flake [fleɪk] n. 薄片，白点，（由白点引起的）开裂
intergranular [ˌɪntəˈgrænjʊlə] adj. 晶粒间的，粒间的
intergranular stress corrosion cracking (IGSCC) 晶间应力腐蚀裂纹
intergranular attack (IGA) 晶间腐蚀
corrosion cell 腐蚀电池
embrittlement [emˈbrɪtlmənt] n. 变脆，脆化，发脆，脆变

hydrogen embrittlement 氢脆（化），氢脆变（钢的）
ion *n.* 离子
delamination [diˌlæməˈneɪʃən] *n.* 分层，剥离
yield stress 屈服应力
brittle failure 脆断，脆性破
freezing point 冰点，凝固点
thermal cycling 热循环
crazing *n.* 细裂纹，银纹，龟裂，网印，网纹，锭模龟裂痕（钢锭缺陷）

creep cracking 蠕变裂纹
brittle fracture 脆性断裂
plastic deformation 塑性变形
burring *n.* 去飞翅
grinding medium 磨料，研磨剂
fatigue resistance 耐［抗］疲劳性
surface finish 表面处理，表面抛光，表面精整

A Sheet Work Manual

Visual Inspection Procedure

1. Scope

This procedure is used for visual examination on the power piping which is fabricated in according to B31.1.

2. Examiner

All personnel performing visual examination in JDB shall be qualified according to JDB/ASME-12.

3. Examination method

(1) Direct visual examination. Direct visual examination may usually be made when access is sufficient to place the eye within 24 in. (600 mm) of the surface to be examined and at an angle not less than 30 deg. to the surface to be examined. Mirrors may be used to improve the angle of vision, and aids such as a magnifying lens may be used to assist examinations. Illumination (natural or supplemental white light) for the specific part, component, vessel, or section thereof being examined is required. The minimum light intensity at examination surface/site shall be 100 footcandles (1 000 lx). The light source, technique used, and light level verification is required to be demonstrated one time, documented, and maintained on file.

(2) Remote visual examination. In some cases, remote visual examination may have to be substituted for direct examination. Remote visual examination may use visual aids such as mirrors, telescopes, borescopes, fiber optics, cameras or other suitable instruments. Such systems shall have a resolution capability at least equivalent to that obtainable by direct visual observation.

4. Piping visual examination

Visual examination shall be performed, as necessary, during the fabrication and erection of piping components to provide verification that the design and WPS requirements are being met. In addition, visual examination shall be performed to verify that all completed welds in pipe and piping components comply with the acceptance standards specified in 3.1 below or with the limitations on imperfections specified in the material specification under which the pipe or component was furnished.

The following indications are unacceptable:

Cracks—external surface;

Undercut on surface which is greater than 1/32 in. (1.0 mm) deep;

Weld reinforcement greater than specified in Table 4–3.

Item 05 Radiographic Testing

Learning Objectives

1. Knowledge objectives
(1) To grasp the words, related terms and abbreviations about RT.
(2) To grasp the classification about RT system.
(3) To know the Instruments and Equipment of RT system.
(4) To know the testing procedure of RT system.

2. Competence objectives
(1) To be able to read and understand frequently used & complex sentence patterns, capitalized English materials and obtain key information quickly.
(2) To be able to communicate with English speakers about the topic freely.
(3) To be able to fill in the job cards in English.

3. Quality objectives
(1) To be able to self-study with the help of aviation dictionaries, the Internet and other resources.
(2) To do a good job of detection of safety protection.

5.1 Theory of Radiographic Testing

Radiography (Fig. 5-1) often referred to as RT or X-ray, is a non-destructive testing method that provides detection of flaws located within objects. It is considered a volumetric examination method because it provides information about the internal regions of parts.

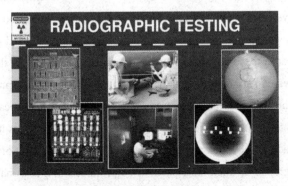

Fig. 5-1 Radiography

■ 〖Point 1〗 History

X-rays were discovered in 1895 by Wilhelm Conrad Roentgen (1845—1923) who was a Professor at Wuerzburg University in Germany. Working with a cathode-ray tube in his laboratory, Roentgen observed a fluorescent glow of crystals on a table near his tube. The tube that Roentgen was working with consisted of a glass envelope (bulb) with positive and negative electrodes encapsulated in it. The air in the tube was evacuated, and when a high voltage was applied, the tube produced a fluorescent glow. Roentgen shielded the tube with heavy black paper, and discovered a green colored fluorescent light generated by a material located a few feet away from the tube.

He concluded that a new type of ray was being emitted from the tube. This ray was capable of passing through the heavy paper covering and exciting the phosphorescent materials in the room. He found that the new ray could pass through most substances casting shadows of solid objects. Roentgen also discovered that the ray could pass through the tissue of humans, but not bones and metal objects. One of Roentgen's first experiments late in 1895 was a film of the hand of his wife, Bertha. It is interesting that the first use of X-rays was for an industrial (not medical) application, as Roentgen produced a radiograph of a set of weights in a box to show his colleagues.

Roentgen's discovery was a scientific bombshell, and was received with extraordinary interest by both scientist and laymen. Scientists everywhere could duplicate his experiment because the cathode tube was very well known during this period. Many scientists dropped other lines of research to pursue the mysterious rays. Newspapers and magazines of the day provided the public with numerous stories, some true, others fanciful, about the properties of the newly discovered rays.

Public fancy was caught by this invisible ray with the ability to pass through solid matter, and in conjunction with a photographic plate, provide a picture of bones and interior body parts. Scientific fancy was captured by the demonstration of a wavelength shorter than light. These generated new possibilities in physics, and for investigating the structure of matter. Much enthusiasm was generated about potential applications of rays as an aid in medicine and surgery. Within a month after the announcement of the discovery, several medical radiographs had been made in Europe and the United States, which were used by surgeons to guide them in their work. In June, 1896, only 6 months after Roentgen announced his discovery, X-rays were being used by battlefield physicians to locate bullets in wounded soldiers.

Prior to 1912, X-rays were used little outside the realms of medicine and dentistry, though some X-ray pictures of metals were produced. The reason that X-rays were not used in industrial application before this date was because the X-ray tubes (the source of the X-rays) broke down under the voltages required to produce rays of satisfactory penetrating power for industrial purposes. However, that changed in 1913 when the high vacuum X-ray tubes designed by Coolidge became available. The high vacuum tubes were an intense and reliable X-ray source, operating at energies up to 100 000 Volts.

In 1922, industrial radiography took another step forward with the advent of the 200 000-Volt X-ray tube that allowed radiographs of thick steel parts to be produced in a reasonable amount of time. In 1931, General Electric Company developed 1 000 000-Volt X-ray generators, providing an effective tool for industrial radiography. That same year, the American Society of Mechanical Engineers (ASME) permitted X-ray approval of fusion welded pressure vessels that further opened the door to industrial acceptance and use.

Shortly after the discovery of X-rays, another form of penetrating rays was discovered. In 1896, French scientist Henri Becquerel discovered natural radioactivity. Many scientists of the period were working with cathode rays, and other scientists were gathering evidence on the theory that the atom could be subdivided. Some of the new research showed that certain types of atoms disintegrate by themselves. It was Henri Becquerel who discovered this phenomenon while investigating the properties of fluorescent minerals. Becquerel was researching the principles of fluorescence, wherein certain minerals glow (fluoresce) when exposed to sunlight. He utilized photographic plates to record this fluorescence.

One of the minerals Becquerel worked with was a uranium compound. On a day when it was too cloudy to expose his samples to direct sunlight, Becquerel stored some of the compound in a drawer with his photographic plates. Later when he developed these plates, he discovered that they were fogged (exhibited exposure to light) . Becquerel questioned what would have caused this fogging. He knew he had wrapped the plates tightly before using them, so the fogging was not due to stray light. In addition, he noticed that only the plates that were in the drawer with the uranium compound were fogged. Becquerel concluded that the uranium compound gave off a type of radiation that could penetrate heavy paper and expose photographic film. Becquerel continued to test samples of uranium compounds and determined that the source of radiation was the element uranium. Bacquerel's discovery was, unlike that of the X-rays, virtually unnoticed by laymen and scientists alike. Relatively few scientists were interested in Becquerel's findings. It was not until the discovery of radium by the Curies two years later that interest in radioactivity became widespread.

While working in France at the time of Becquerel's discovery, Polish scientist Marie Curie became very interested in his work. She suspected that a uranium ore known as pitchblende contained other radioactive elements. Marie and her husband, French scientist Pierre Curie, started looking for these other elements. In 1898, the Curies discovered another radioactive element in pitchblende, and named it "polonium" in honor of Marie Curie's native homeland. Later that year, the Curies discovered another radioactive element which they named radium, or shining element. Both polonium and radium were more radioactive than uranium. Since these discoveries, many other radioactive elements have been discovered or produced.

Radium became the initial industrial gamma ray source. The material allowed castings up to 10 to 12 inches thick to be radiographed. During World War II, industrial radiography grew tremendously as part of the Navy's shipbuilding program. In 1946, man-made gamma ray sources such as cobalt and iridium became available. These new sources were far stronger than radium and were much less expensive. The manmade sources rapidly replaced radium, and use of gamma rays grew quickly in industrial radiography.

This method of testing (Fig. 5–2) can be used on virtually any material, metallic and nonmetallic, since radiation can penetrate all forms of matter. It can be used to inspect castings, forgings or weldments. The results of radiographic testing are most commonly viewed on a radiograph or X-ray which becomes a permanent physical record of the testing. Digital and real-time radiographic techniques are becoming more common more widely used. These allow the images to be viewed and stored electronically.

Simulated rendering of radiographic inspection

X-ray image of part pictured above

Fig. 5-2 Method of Testing

〖Point 2〗 Basic Principle (Fig. 5-3)

Radiography is a non-destructive inspection technique used for materials failure analysis investigation in different industries including oil, gas and aerospace. This technique is an internal examination of materials using X-ray, gamma ray or neutron radiation. Radiography is based on the differential adsorption of the penetrating radiation into the target material. Radiation is directed through the sample to a photography film. Path length of radiation and density of sample are the two main parameters determining the intensity of light passing through the sample and creating contrast on the photographic film. Internal discontinuities such as porosity and cracks can be detected using this technique.

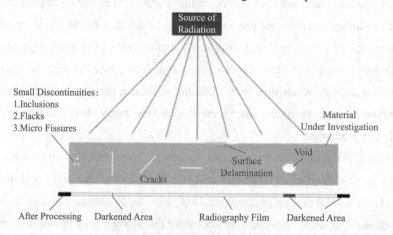

Fig. 5-3 Basic Principle

Radiography or X-ray, is another common way to inspect welds. It uses the concept of differential absorption of penetrating radiation. Specimens will have differences in density, thickness, shapes, sizes or absorption characteristics. Unabsorbed radiation that passes through a part is recorded on film, fluorescent screens or other radiation monitors. Indications of internal and external conditions will appear as variants of black/white/gray contrasts on exposed film, or variants of color on fluorescent screens.

Radiography (Fig. 5-4) is used widely in the examination of castings and weldments, particularly where there is a critical need to ensure freedom from internal flaws.

The radiation used is either X-radiation or gamma radiation (Fig. 5-5). X-radiation is produced electrically in an X-ray tube and is familiar to people since they encounter this with medical or dental procedures. Gamma radiation is emitted by radioactive materials and is utilized more commonly in field radiography since it does not require a power source. Radioactive materials and gamma radiation are also used in medical treatments.

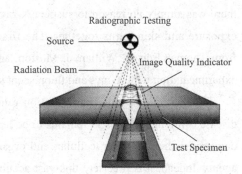

Fig. 5-4 Radiography Ability for Detecting Different Types of Defects and Discontinuities in Materials

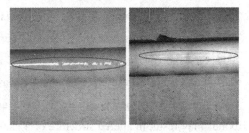

Fig. 5-5 Shows Examples of Weld Tunnels Being Detected By X-ray

Radiographic testing is limited by the overall part configuration and the equipment capabilities available. The relationship between the X-ray beam direction and flaw orientation determines the detectability of flaws. Due to safety considerations, radiography is typically performed in a cabinet or a vault to shield the operator from radiation exposure.

Radiography is one of the more expensive methods of testing. Contributing factors are the radiographic film which contains silver, chemicals used for processing the radiographs, hazardous waste disposal, and licensing issues related to the radiographic equipment.

■ 〖Point 3〗 Health Concerns

The science of radiation protection, or "health physics" as it is more properly called, grew out of the parallel discoveries of X-rays and radioactivity in the closing years of the 19th century. Experimenters, physicians, laymen, and physicists alike set up X-ray generating apparatuses and proceeded about their labors with a lack of concern regarding potential dangers. Such a lack of concern is quite understandable, for there was nothing in previous experience to suggest that X-rays would in any way be hazardous. Indeed, the opposite was the case, for who would suspect that a ray similar to light but unseen, unfelt, or otherwise undetectable by the senses would be damaging to a person? More likely, or so it seemed to some, X-rays could be beneficial for the body.

Inevitably, the widespread and unrestrained use of X-rays led to serious injuries. Often injuries were not attributed to X-ray exposure, in part because of the slow onset of symptoms, and because there was simply no reason to suspect X-rays as the cause. Some early experimenters did tie X-ray exposure and skin burns together. The first warning of possible adverse effects of X-rays came from Thomas Edison, William J. Morton, and Nikola Tesla who each reported eye irritations from experimentation with X-rays and fluorescent substances.

Today, it can be said that radiation ranks among the most thoroughly investigated causes of disease. Although much still remains to be learned, more is known about the mechanisms of radiation damage on the molecular, cellular, and organ system than is known for most other health stressing agents. Indeed, it is precisely this vast accumulation of quantitative dose-response data that enables health physicists to specify radiation levels so that medical, scientific, and industrial uses of radiation may continue at levels of risk no greater than, and frequently less than, the levels of risk associated with any other technology.

X-rays and Gamma rays are electromagnetic radiation of exactly the same nature as light, but of much shorter wavelength. Wavelength of visible light is on the order of 6 000 angstroms while the wavelength of X-rays is in the range of one angstrom and that of gamma rays is 0.000 1 angstrom. This very short wavelength is what gives X-rays and gamma rays their power to penetrate materials that light cannot. These electromagnetic waves are of a high energy level and can break chemical bonds in materials they penetrate. If the irradiated matter is living tissue, the breaking of chemical bonds may result in altered structure or a change in the function of cells. Early exposures to radiation resulted in the loss of limbs and even lives. Men and women researchers collected and documented information on the interaction of radiation and the human body. This early information helped science understand how electromagnetic radiation interacts with living tissue. Unfortunately, much of this information was collected at great personal expense.

■〖Point 4〗 Application

Radiography can be used for inspection of welds, castings and wrought materials, in metallic and non-metallic materials.

When the film (Fig. 5-6) is processed a negative is produced. The thin areas of an object will be darker than the thicker areas, therefore most weld defects will show up dark in relation to the surrounding areas.

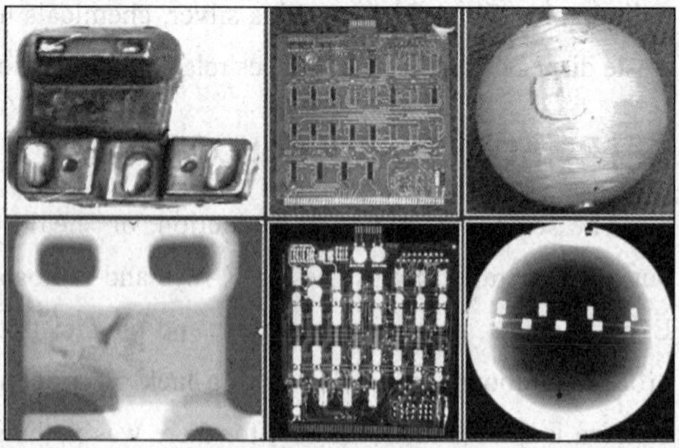

Radiography advantages:

• Virtually no surface preparation is required.

Fig. 5-6　Film

• Volumetric: both surface and sub-surface indications can be found.
• Suitable for inspection of assembled components.
• Suitable for in site and on-stream inspections.

- Suitable for many different material types.
- Permanent record deliverables (film or digital file).

Radiography limitations:

- Hazardous ionising radiation requires creation of safety perimeter.
- Relatively expensive equipment.
- Relatively slow inspection process.
- Sensitive to flaw orientation.
- Usualy not possible to determine depth of indications.
- Two-sided access to test object is required.

Put into Practice

1. Translate the ray detection schematic in Fig. 5-7.

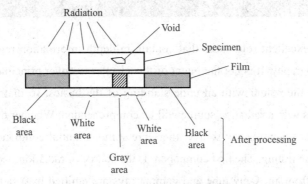

Fig. 5-7 Ray Detection Schematic

2. Read Table 5-1 to analyze the advantages and disadvantages of ray detection and the scope of application.

Table 5-1 The Advantages and Disadvantages of Ray Detection

Monitoring/inspection type	Application	Advantages	Disadvantages
· Radiography	· Inclusions · Cracks · Porosity · Corrosion · Debris · Lack of fusion · Lack of penetration · Leak paths	· Sensitive to finding discontinuities throughout the volume of materials · Easily understood permanent record · Full volumetric examination · Portability	· Radiation hazard · Relatively expensive · Long set-up time · Necessary access to both sides of specimen · Depth of indication not shown · High degree of skill required for technique and interpretation · Lack of sensitivity to fine cracks and lack of penetration

3. Read Fig. 5-8 to show how to reduce radiation emissions?

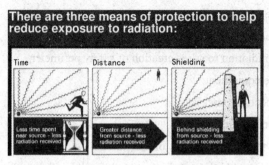

Fig. 5-8 How to Reduce Radiation Emissions

5.2 Physics of Radiography

There are many excellent references that contain in-depth information regarding the principles of radiation and radiography. It is not the intent of this chapter to cover principles and theory in great depth but to provide the reader with an understanding of the basics, so that the entire process of producing radiographs with a radiation source will be comprehended. Whether the radiation is emitted from an X-ray tube or a gamma ray source, there are some essential components that apply to the process of radiographic testing. The first component is the source of radiation.

X-rays emanate from an X-ray tube and gamma rays are emitted by a radioactive isotope. The second component is the test specimen that is to be examined. The third includes the development of the technique. The fourth involves the taking of the radiograph and the processing of the film. The final component, which is extremely critical, is the interpretation of the radiographic image.

■ 〖Point 1〗 Characteristics of Radiation

There are certain unique characteristics relative to radiation that must be understood in order to realize the physics and variables involved with producing a radiograph with the use of a radiation source. These characteristics apply to both X-radiation and gamma radiation. Recall that the only difference between X-radiation and gamma radiation is their origin. X-rays are produced by an X-ray tube and gamma rays come from a radioactive source that is disintegrating. The following characteristics apply to the radiation that will be used in the non-destructive examination of materials.

X- and gamma rays are part of what scientists refer to as the electromagnetic spectrum. They are waveforms that are part of a family in which some of the relatives are very familiar to us, such as light rays, infrared heat rays, and radio waves. However, X- and gamma rays cannot been seen, felt or heard. In other words, our normal senses cannot detect them. Since X- and gamma rays have no mass and no electrical charge, they are not influenced by electrical and magnetic fields and will travel in straight lines. Continued research over the years since Roentgen's discovery indicated that the radiation possesses a dual character. Acting somewhat like a particle at times and like a wave at other times. The name that has been given to the small "packets" of energy with these characteristics is "photon". It is said that the radiation photon is a wave that is both electric and magnetic in nature. Electromagnetic radiation has also been described in terms of a stream of photons (massless particles) each traveling in a wave-like pattern and moving at the speed of light.

Fig. 5-9 shows the electromagnetic spectrum. Notice the changes in wavelengths of the various wave forms.

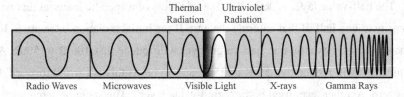

Fig. 5-9   Electromagnetic Spectrum

Every point across the spectrum represents a wave form of differing wavelength. It should be noted that the lines between the groupings are not precise, and that each group phases into the next.

Radiation is absorbed and scattered by material. There are four common absorption processes that influence the amount of radiation that passes through a part. The four absorption processes are:

(1) Photoelectric effect;

(2) Rayleigh scattering;

(3) Compton scattering;

(4) Pair production.

The photoelectric effect is that process in which a photon of low radiation energy (less than 500 kV) transfers all of its energy to an electron in some shell of the material atom. The energy may simply move an electron from one shell to another, or if there is energy above that required to interact with the orbital electron in the material, it will impart kinetic energy and the electron will be ejected from the atom. Another name for Rayleigh scattering is "coherent scattering" and it is a direct interaction between the photon and the orbital electrons in the atom of the material. However, in this case the photon is deflected without any change or reduction of the kinetic energy of the photon or of the energy of the material atoms.

In this process, there are no electrons released from the atom. It is estimated that rayleigh scattering accounts for no more than about 20% of the total attenuation. Compton scattering is a direct interaction between photons in the 0.1—3.0 MeV energy range and an orbital electron. In this case, when the electron is ejected from the material atom, only a portion of the kinetic energy of the photon is used. The photon then scatters in a different direction than the direction from which it came and actually emerges with a reduced energy and, therefore, a lower wavelength. Compton scattering varies with the atomic number of the material and varies, roughly, inversely with the energy of the photon.

Pair production is an attenuation process that results in the creation of two 0.51 MeV photons (as a result of annihilation of the electron-positive pair) of scattered radiation for each high-energy incident photon that is at or above an energy of 1.02 MeV. The two 0.51 MeV photons travel in different directions, causing the production of electromagnetic radiation through interaction with other material particles. Energies exceeding 1.02 MeV result in additional kinetic energy being applied to the pair of particles. Total absorption, therefore, is the combined sums of the four different types of absorption.

〖Point 2〗 Radiation Penetrates

The variables relating to the penetration of the radiation can be expressed with the term "half-value layer". The half-value layer is defined as the thickness of a specific material that will reduce the radiation intensity to one-half of that entering the part. If the initial radiation intensity is 100 roentgens (100 R), a material that is exactly one half-value layer will reduce that 100 R to 50 R. Another half-value layer thickness will reduce that 50 R to 25 R, and so on. If this is carried forward, the radiation never reaches zero. Another term used is the "tenth-value layer". The same principle as the half-value layer applies, except that the thickness is somewhat greater and reduces the initial radiation to one-tenth on the opposite side. The factors involved with these half- and tenth-value layers (Table 5-2) are:

(1) Energy. The higher the energy, the thicker the half- or tenth-value layers. This is supported by the fact that the higher energy radiation produces shorter wavelength radiation, resulting in better penetration. The half-value and tenth-value layers are especially useful for calculating the thickness of shielding when designing an enclosure or room for radiography.

(2) Material type. The greater the material density, the thinner the half-value layer. Material that is low in density, such as aluminum or titanium, will allow more radiation to pass through, and there will be less absorption and scattering. Materials with higher density, such as steel and lead, provide a much greater chance of interaction because of their higher atomic number. With higher-density materials there will be more absorption and the half-value layer thickness will be considerably less than with the lower density materials.

(3) Thickness. As mentioned, the half-value layer thickness is specific for a given material and energy. As the thickness of the material increases, the amount of absorption and scatter increases and the amount of radiation that passes through that thickness.

Table 5-2 Half- and Tenth-Value Layers

X-ray/kV	Half-value		Tenth-value		Half-value	
	Lead /in	Concrete /in	Lead /in	Concrete /in	X-ray	Steel /in
50	0.002	0.2	0.007	0.66	120 kV	0.10
70	0.007	0.5	0.023	1.65	150 kV	0.14
100	0.009	0.7	0.028	2.31	200 kV	0.20
125	0.011	0.8	0.035	2.64	250 kV	0.25
150	0.012	0.9	0.039	2.79	400 kV	0.35

continued

X-ray/kV	Half-value		Tenth-value		X-ray	Half-value
	Lead /in	Concrete /in	Lead /in	Concrete /in		Steel /in
200	0.020	1.0	0.065	3.30	1 MeV	0.60
250	0.032	1.1	0.104	3.63	2 MeV	0.80
300	0.059	1.2	0.195	3.96	4 MeV	1.00
400	0.087	1.3	0.286	4.29	6 MeV	1.15
1 000	0.315	1.8	1.04	5.94	10 MeV	1.25
2 000	0.393	2.45	1.299	8.60	16 MeV+	1.30
Isotopes						
Iridium 192	0.190	1.9	0.640	6.20		0.60
Cesium 137	0.250	2.1	0.840	7.10		0.68
Cobalt 60	0.490	2.6	1.620	8.60		0.87

〖Point 3〗 Other

1. Radiation travels in straight lines and at the speed of light

The speed of light is 186 300 miles per second or 299 800 km per second.

2. Radiation exhibits energy

On the electromagnetic spectrum, the wavelengths of X- and gamma rays are much shorter than that of visible light. Recall that as the radiation energy increases, shorter wavelengths are produced that provide greater penetration through the material being examined.

3. Radiation ionizes

Matter Ionization is a change in the electrical nature or characteristics of matter. Electrons are displaced or knocked out of orbit, thereby changing the electrical balance. This, coincidentally, is what causes the greatest concern to humans. When radiation passes through living tissue, the cells are affected—electrically changed—and damage will result.

4. Radiation is not particulate

Radiation has no mass, and even though X- and gamma rays behave like particles, they are actually weightless. To describe the effect of radiation as it interacts with matter, the radiation is sometimes referred to as photons, which is another way of saying high-speed energy traveling at the speed of light that behaves like particles; but in fact, there are no particles in X- or gamma radiation.

5. Radiation has no electrical charge

X- and gamma radiation are not affected by either strong electrical or magnetic fields.

6. X- and gamma radiation cannot be focused

If X- or gamma radiation is directed toward a glass lens, the radiation would not focus like light does. In fact, the radiation passing through that lens would be absorbed to a greater extent in the thicker portion of the lens and more radiation would pass through the thinner regions of the lens.

7. Humans cannot sense X- or gamma radiation

Radiation cannot be seen, tasted, or felt. It has no odor and if humans are subjected to radiation exposure, the effects are not realized for a period of time. In other words, the body goes through a latent period before the harmful effects of the radiation exposure are evident.

8. Radiation causes fluorescence in some materials

Some minerals and salts will fluoresce when subjected to radiation. Fluoroscopy utilizes a fluorescent screen that converts some of the radiation to light. When used in radiography, the light that is emitted from the fluorescent screen creates a much higher exposure effect on the film that is adjacent to it, thereby significantly reducing exposure.

9. Intensity of radiation

Radiation is expressed in roentgens per hour (R/hr), roentgens per minute (R/min), or mill roentgens per hour (mR/hr) . The intensity decreases with distance. This is a function of the inverse square law, which states that the intensity of the radiation varies inversely with the square of the distance. As the distance from the radiation source is doubled, the intensity is decreased to one-fourth.

10. Radiation produces chemical changes in exposed film

The film's emulsion is chemically changed through ionization as radiation interacts with it. If it weren't for this characteristic, it would not be possible to produce a radiographic image on film.

Put into Practice

1. Explain what the effects are in each of Fig. 5-10.

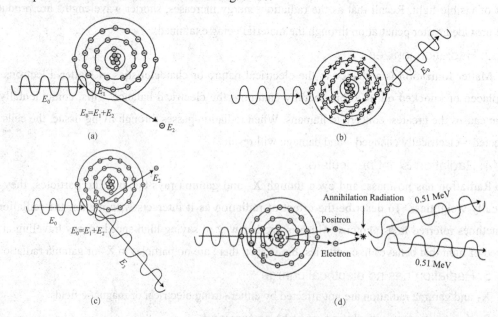

Fig. 5-10　Effects

(a) Photoelectric Effect; (b) Rayleigh or Coherent Scattering;
(c) Compton Scattering; (d) Pair Production

2. Look at Fig. 5-11 to show what the effect of high and low energy on the half-price layer is.

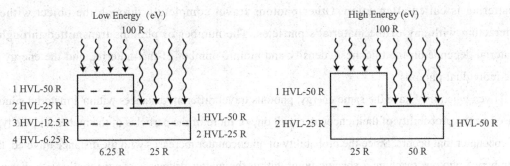

Fig. 5-11　The Effect of High and Low Energy on the Half-price Layer

5.3　Principles of X-radiography

〖Point 1〗 Newton's Inverse Square Law

Any point source which spreads its influence equally in all directions without a limit to its range will obey the inverse square law. This comes from strictly geometrical considerations. The intensity of the influence at any given radius (r) is the source strength divided by the area of the sphere. Being strictly geometric in its origin, the inverse square law applies to diverse phenomena. Point sources of gravitational force, electric field, light, sound and radiation obey the inverse square law (Fig. 5-12).

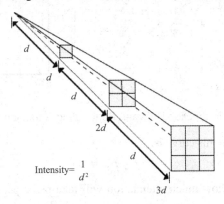

Fig. 5-12　Inverse Square Law

As one of the fields which obey the general inverse square law, a point radiation source can be characterized by the diagram above whether you are talking about Roentgens, rads, or rems. All measures of exposure will drop off by the inverse square law.

〖Point 2〗 Interaction Between Penetrating Radiation and Matter

When X-rays or gamma rays are directed into an object, some of the photons interact with the particles of the matter and their energy can be absorbed or scattered. This absorption and scattering is called attenuation. Other photons travel completely through the object without interacting with any of the material's particles. The number of photons transmitted through a material depends on the thickness, density and atomic number of the material, and the energy of the individual photons.

Even when they have the same energy, photons travel different distances within a material simply based on the probability of their encounter with one or more of the particles of the matter and the type of encounter that occurs. Since the probability of an encounter increases with the distance traveled, the number of photons reaching a specific point within the matter decreases exponentially with distance traveled. As shown in Fig. 5-13, if 1 000 photons are aimed at ten 1 cm layers of a material and there is a 10% chance of a photon being attenuated in this layer, then there will be 100 photons attenuated. This leave 900 photons to travel into the next layer where 10% of these photons will be attenuated. By continuing this progression, the exponential shape of the curve becomes apparent.

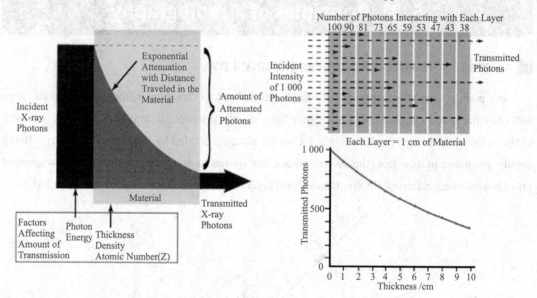

Fig. 5-13   Interaction Between Penetrating Radiation and Matter

The formula that describes this curve is:

$$I = I_0 e^{-\mu x}$$

The factor that indicates how much attenuation will take place per cm (10% in this example) is known as the linear attenuation coefficient m. The above equation and the linear attenuation coefficient will be discussed in more detail on the following page.

■ 〖Point 3〗 Transmitted Intensity and Linear Attenuation Coefficient

For a narrow beam of mono-energetic photons, the change in X-ray beam intensity at some distance in a material can be expressed in the form of an equation as:

$$dI(\chi) = -I(\chi) \cdot n \cdot \sigma \cdot dx$$

Where: dI—the change in intensity;

I—the initial intensity;

n—the number of atoms/cm^3;

σ—a proportionality constant that reflects the total probability of a photon being scattered or absorbed;

dx—the incremental thickness of material traversed.

When this equation is integrated, it becomes:

$$I = I_0 e^{-n\sigma x}$$

The number of atoms/cm^3 (n) and the proportionality constant (S) are usually combined to yield the linear attenuation coefficient (m). Therefore, the equation becomes:

$$I = I_0 e^{-\mu x}$$

Where: I—the intensity of photons transmitted across some distance x;

I_0—the initial intensity of photons;

S—a proportionality constant that reflects the total probability of a photon being scattered or absorbed;

μ—the linear attenuation coefficient;

x—distance traveled.

The Linear Attenuation Coefficient (μ)

The linear attenuation coefficient (μ) describes the fraction of a beam of X-rays or gamma rays that is absorbed or scattered per unit thickness of the absorber. This value basically accounts for the number of atoms in a cubic cm volume of material and the probability of a photon being scattered or absorbed from the nucleus or an electron of one of these atoms.

Using the transmitted intensity equation above, linear attenuation coefficients can be used to make a number of calculations. These include:

The intensity of the energy transmitted through a material when the incident X-ray intensity, the material and the material thickness are known.

The intensity of the incident X-ray energy when the transmitted X-ray intensity, material, and material thickness are known.

The thickness of the material when the incident and transmitted intensity, and the material are known.

The material can be determined from the value of μ when the incident and transmitted intensity, and the material thickness are known.

One use of linear attenuation coefficients is for selecting a radiation energy that will produce the most contrast between particular materials in a radiograph. Say, for example, that it is necessary to detect tungsten inclusions in iron. It can be seen from Fig. 5–14 of linear attenuation coefficients versus radiation energy, that the maximum separation between the tungsten and iron curves occurs at around 100 keV. At this energy the difference in attenuation between the two materials is the greatest so the radiographic contrast will be maximized.

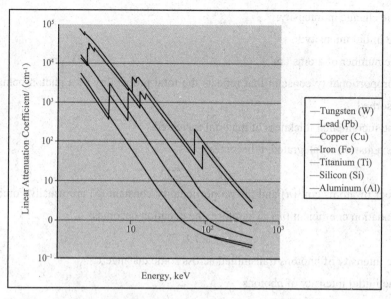

Fig. 5-14 Linear Attenuation Coefficients

〖Point 4〗 Half-value Layer

The thickness of any given material where 50% of the incident energy has been attenuated is know as the half-value layer (HVL) (Fig. 5–15). The HVL is expressed in units of distance (mm or cm). Like the attenuation coefficient, it is photon energy dependant. Increasing the penetrating energy of a stream of photons will result in an increase in a material's HVL.

The HVL is inversely proportional to the attenuation coefficient. If an incident energy of 1 and a transmitted energy is 0.5 is plugged into the equation introduced on the preceding page, it can be seen that the HVL multiplied by m must equal 0.693.

Therefore, the HVL and m are related as follows:

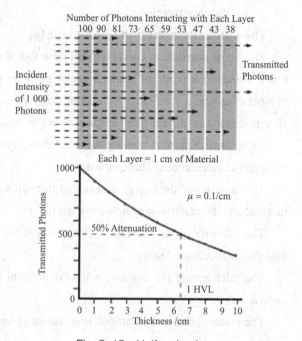

Fig. 5-15 Half-value Layer

$$HVL = \frac{0.693}{\mu}$$

The HVL is often used in radiography simply because it is easier to remember values and perform simple calculations. In a shielding calculation, such as illustrated to the below (Fig. 5–16), it can be seen that if the thickness of one HVL is known, it is possible to quickly determine how much material is needed to reduce the intensity to less than 1%.

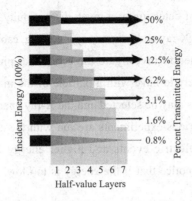

Fig. 5-16 Percent Transmitted Energy

〔Point 5〕 Sources of Attenuation

The attenuation that results due to the interaction between penetrating radiation and matter is not a simple process. A single interaction event between a primary X-ray photon and a particle of matter does not usually result in the photon changing to some other form of energy and effectively disappearing. Several interaction events are usually involved and the total attenuation is the sum of the attenuation due to different types of interactions. These interactions include the photoelectric effect, scattering and pair production. The Fig. 5-17 shows an approximation of the total absorption coefficient (μ), in gray, for iron plotted as a function of radiation energy. The four radiation-matter interactions that contribute to the total absorption are shown in black. The four types of interactions are photoelectric (PE), Compton scattering (C), pair production (PP) and Thomson or Rayleigh scattering (R). Since most industrial radiography is done in the 0.1 to 1.5 MeV range, it can be seen from the plot that photoelectric and Compton scattering account for the majority of attenuation encountered.

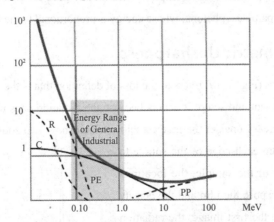

Fig. 5-17 Absorption Characteristics

Absorption characteristics will increase or decrease as the energy of the X-ray is increased or decreased. Since attenuation characteristics of materials are important in the development of contrast in a radiograph, an understanding of the relationship between material thickness, absorption properties, and photon energy is fundamental to producing a quality radiograph. A radiograph with higher contrast will provide greater probability of detection of a given discontinuity. An understanding of absorption is also necessary when designing X-ray and gamma ray shielding, cabinets or exposure vaults.

The applet below can be used to investigate the effect that photon energy has on the type of interaction that the photon is likely to have with a particle of the material (shown in gray) . Various materials and material thicknesses may be selected and the X-ray energy can be set to produce a range from 1 to 199 keV. Notice as various experiments are run with the applets that low energy radiation produces predominately photoelectric events and higher energy X-rays produce predominately compton scattering events. Also notice that if the energy is too low, none of the radiation penetrates the material.

■ 〖Point 6〗 Compton Scattering

As mentioned on the previous page, compton scattering (Fig. 5-18) occurs when the incident X-ray photon is deflected from its original path by an interaction with an electron. The electron is ejected from its orbital position and the X-ray photon loses energy because of the interaction but continues to travel through the material along an altered path. Energy and momentum are conserved in this process. The energy shift depends on the angle of scattering and not on the nature of the scattering medium. Since the scattered X-ray photon has less energy, it has a longer wavelength and less penetrating than the incident photon.

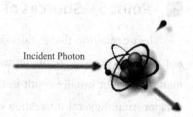

Fig. 5-18 Compton Scattering

Compton effect was first observed by Arthur Compton in 1923 and this discovery led to his award of the 1927 Nobel Prize in Physics. The discovery is important because it demonstrates that light cannot be explained purely as a wave phenomenon. Compton's work convinced the scientific community that light can behave as a stream of particles (photons) whose energy is proportional to the frequency.

■ 〖Point 7〗 Geometric Unsharpness

Geometric unsharpness (Fig. 5-19) refers to the loss of definition that is the result of geometric factors of the radiographic equipment and setup. It occurs because the radiation does not originate from a single point but rather over an area. Consider the images below which show two sources of different sizes, the paths of the radiation from each edge of the source to each edge of the feature of the sample, the locations where this radiation will expose the film and the density profile across the film. In the first image, the radiation originates at a very small source. Since all of the radiation originates from basically the same point, very little geometric unsharpness is produced in the image. In the second image, the source size is larger and the different paths that the rays of radiation can take from their point

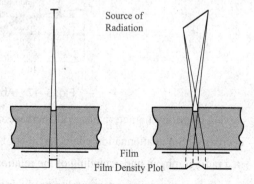

Fig. 5-19 Geometric Unsharpness

of origin in the source causes the edges of the notch to be less defined.

The three factors controlling unsharpness are source size, source to object distance, and object to detector distance. The source size is obtained by referencing manufacturers specifications for a given X-ray or gamma ray source. Industrial X-ray tubes often have focal spot sizes of 1.5 mm squared but microfocus systems have spot sizes in the 30 micron range. As the source size decreases, the geometric unsharpness also decreases. For a given size source, the unsharpness can also be decreased by increasing the source to object distance, but this comes with a reduction in radiation intensity.

Put into Practice

1. Translate Table 5–3.

Table 5–3   Gamma Ray

Source	Symbol	Atomic number	Atomic Weight	Isotope	Half-life	Energy	Emissivity
Cesium	Cs	55	132.91	137	30 years	0.66 meV	4. 2 R/C hr@1 ft
Cobalt	Co	27	58.9	60	5.3 years	1.17 meV 1.33 meV	14 .5 R/C hr@1 ft
Iridium	Ir	77	192.2	192	75 days	0.61 meV 0.21 meV	5. 9 R/C hr@1 ft
Radium	Ra	88	226	226	1 602 years	2.2 meV 0.24 meV	9. 0 R/C hr@1 ft

2. Look at Fig. 5–20 to compare the penetration performance of three different rays.

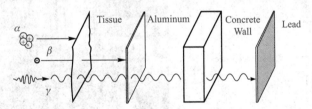

Fig. 5–20   The Penetration Performance

3. Translate Fig. 5–21.

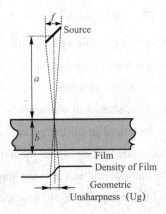

Fig. 5–21   RT

5.4 Equipment & Materials

■ 〔Point 1〕 X-ray Generators

The major components of an X-ray generator are the tube, the high voltage generator, the control console, and the cooling system. As discussed earlier in this material, X-rays are generated by directing a stream of high speed electrons at a target material such as tungsten, which has a high atomic number. When the electrons are slowed or stopped by the interaction with the atomic particles of the target, X-radiation is produced. This is accomplished in an X-ray tube such as the one shown here. The X-ray tube is one of the components of an X-ray generator and tubes come a variety of shapes and sizes. Fig. 5–22 shows a portion of the Roentgen tube collection of Grzegorz Jezierski, a professor at Opole University of Technology.

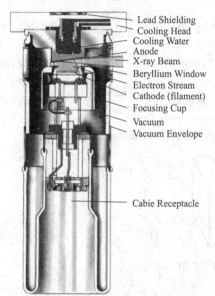

Fig. 5–22   X-ray Generator

The tube cathode (filament) is heated with a low-voltage current of a few amps. The filament heats up and the electrons in the wire become loosely held. A large electrical potential is created between the cathode and the anode by the high-voltage generator. Electrons that break free of the cathode are strongly attracted to the anode target. The stream of electrons between the cathode and the anode is the tube current. The tube current is measured in milliamps and is controlled by regulating the low-voltage, heating current applied to the cathode. The higher the temperature of the filament, the larger the number of electrons that leave the cathode and travel to the anode. The milliamp or current setting on the control console regulates the filament temperature, which relates to the intensity of the X-ray output.

The high-voltage between the cathode and the anode affects the speed at which the electrons travel and strike the anode. The higher the kilovoltage, the more speed and, therefore, energy the electrons have when they strike the anode. Electrons striking with more energy results in X-rays with more penetrating power. The high-voltage potential is measured in kilovolts, and this is controlled with the voltage or kilovoltage control on the control console. An increase in the kilovoltage will also result in an increase in the intensity of the radiation.

A focusing cup is used to concentrate the stream of electrons to a small area of the target called the focal spot. The focal spot size is an important factor in the system's ability to produce a sharp image. See the information on image resolution and geometric unsharpness for more information on the effect of the focal spot size. Much of the energy applied to the tube is transformed into heat at the focal spot of the anode. As mentioned above, the anode target is commonly made from tungsten, which has a high melting point in addition to a high atomic number. However, cooling of the anode by active or passive means is necessary. Water or oil recirculating systems are often used to cool tubes. Some low power tubes are cooled simply with the use of thermally conductive materials and heat radiating fins.

It should also be noted that in order to prevent the cathode from burning up and to prevent arcing between the anode and the cathode, all of the oxygen is removed from the tube by pulling a vacuum. Some systems have external vacuum pumps to remove any oxygen that may have leaked into the tube. However, most industrial X-ray tubes simply require a warm-up procedure to be followed. This warm-up procedure carefully raises the tube current and voltage to slowly burn any of the available oxygen before the tube is operated at high power.

The other important component of an X-ray generating system is the control console (Fig. 5–23). Consoles typically have a keyed lock to prevent unauthorized use of the system. They will have a button to start the generation of X-rays and a button to manually stop the generation of X-rays. The three main adjustable controls regulate the tube voltage in kilovolts, the tube amperage in milliamps, and the exposure time in minutes and seconds. Some systems also have a switch to change the focal spot size of the tube.

Image Courtesy of Yxlon International   Image Courtesy of LORAD Industrial Imaging

Fig. 5–23   X-ray Generator Options

Kilovoltage

X-ray generators come in a large variety of sizes and configurations. There are stationary units that are intended for use in lab or production environments and portable systems that can be easily moved to the job site. Systems are available in a wide range of energy levels. When inspecting large steel or heavy metal components, systems capable of producing millions of electron volts may be necessary to penetrate the full thickness of the material. Alternately, small, lightweight components may only require a system capable of producing only a few tens of kilovolts.

Focal Spot Size

Another important consideration is the focal spot size of the tube since this factors into the geometric unsharpness of the image produced. Generally, the smaller the spot size the better. But as the electron stream is focused to a smaller area, the power of the tube must be reduced to prevent overheating at the tube anode. Therefore, the focal spot size becomes a tradeoff of resolving capability and power. Generators can be classified as a conventional, minifocus and microfocus system. Conventional units have focal-spots larger than about 0.5 mm, minifocus units have focal-spots ranging from 50 to 500 microns (0.05 to 0.5 mm), and microfocus systems have focal-spots smaller than 50 microns. Smaller spot sizes are especially advantageous in instances where the magnification of an object or region of an object is necessary. The cost of a system typically increases as the spot size decreases and some microfocus tubes exceed $100 000. Some manufacturers combine two filaments of different sizes to make a dual-focus tube. This usually involves a conventional and a minifocus spot-size and adds flexibility to the system.

AC and Constant Potential Systems

AC X-ray systems supply the tube with sinusoidal varying alternating current. They produce X-rays only during one half of the 1/60th second cycle. This produces bursts of radiation rather than a constant stream. Additionally, the voltage changes over the cycle and the X-ray energy varies as the voltage ramps up and then back down. Only a portion of the radiation is useable and low energy radiation must usually be filtered out. Constant potential generators rectify the AC wall current and supply the tube with DC current. This results in a constant stream of relatively consistent radiation. Most newer systems now use constant potential generators.

Flash X-ray Generators

Flash X-ray generators produce short, intense bursts of radiation. These systems are useful when examining objects in rapid motion or when studying transient events such as the tripping of an electrical breaker. In these type of situations, high-speed video is used to rapidly capture images from an image intensifier or other real-time detector. Since the exposure time for each image is very short, a high level of radiation intensity is needed in order to get a usable output from the detector. To prevent the imaging system from becoming saturated from a continuous exposure high intensity radiation, the generator supplies microsecond bursts of radiation. The tubes of these X-ray generators do not have a heated filament but instead electrons are pulled from the cathode by the strong electrical potential between the cathode and the anode. This process is known as field emission or cold emission and it is capable of producing electron currents in the thousands of amperes.

【Point 2】 Radiographic Film

X-ray films for general radiography consist of an emulsion-gelatin containing radiation sensitive silver halide crystals, such as silver bromide or silver chloride, and a flexible, transparent, blue-tinted base. The emulsion is different from those used in other types of photography films to account for the distinct characteristics of gamma rays and X-rays, but X-ray films are sensitive to light. Usually, the emulsion is coated on both sides of the base in layers about 0.000 5 inch thick. Putting emulsion on both sides of the base doubles the amount of radiation-sensitive silver halide, and thus increases the film speed. The emulsion layers are thin enough so developing, fixing, and drying can be accomplished in a reasonable time. A few of the films used for radiography only have emulsion on one side which produces the greatest detail in the image.

When X-rays, gamma rays, or light strike the grains of the sensitive silver halide in the emulsion, some of the Br^- ions are liberated and captured by the Ag^+ ions. This change is of such a small nature that it cannot be detected by ordinary physical methods and is called a "latent (hidden) image". However, the exposed grains are now more sensitive to the reduction process when exposed to a chemical solution (developer), and the reaction results in the formation of black, metallic silver. It is this silver, suspended in the gelatin on both sides of the base, that creates an image. See the page on film processing for additional information.

The selection of a film (Fig. 5-24) when radiographing any particular component depends on a number of different factors. Listed below are some of the factors that must be considered when selecting a film and developing a radiographic technique. Composition, shape, and size of the part being examined and, in some cases, its weight and location.

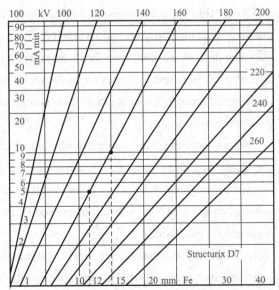

Fig. 5-24 Film Selection

Type of radiation used, whether X-rays from an X-ray generator or gamma rays from a radioactive source.

Kilovoltages available with the X-ray equipment or the intensity of the gamma radiation.

Relative importance of high radiographic detail or quick and economical results.

Selecting the proper film and developing the optimal radiographic technique usually involves arriving at a balance between a number of opposing factors. For example, if high resolution and contrast sensitivity is of overall importance, a slower and finer grained film should be used in place of a faster film.

Film Packaging (Fig. 5-25)

Radiographic film can be purchased in a number of different packaging options. The most basic form is as individual sheets in a box. In preparation for use, each sheet must be loaded into a cassette or film holder in the darkroom to protect it from exposure to light. The sheets are available in a variety of sizes and can be purchased with or without interleaving paper.

Fig. 5-25 Packaging

Interleaved packages have a layer of paper that separates each piece of film. The interleaving paper is removed before the film is loaded into the film holder. Many users find the interleaving paper useful in separating the sheets of film and offer some protection against scratches and dirt during handling.

Industrial X-ray films are also available in a form in which each sheet is enclosed in a light-tight envelope. The film can be exposed from either side without removing it from the protective packaging. A rip strip makes it easy to remove the film in the darkroom for processing. This form of packaging has the advantage of eliminating the process of loading the film holders in the darkroom. The film is completely protected from finger marks and dirt until the time the film is removed from the envelope for processing.

Film Handling

X-ray film should always be handled carefully to avoid physical strains, such as pressure, creasing, buckling, friction, etc. Whenever films are loaded in semi-flexible holders and external clamping devices are used, care should be taken to be sure pressure is uniform. If a film holder bears against a few high spots, such as on an unground weld, the pressure may be great enough to produce desensitized areas in the radiograph. This precaution is particularly important when using envelope-packed films.

Marks resulting from contact with fingers that are moist or contaminated with processing chemicals, as well as crimp marks, are avoided if large films are always grasped by the edges and allowed to hang free. A supply of clean towels should be kept close at hand as an incentive to dry the hands often and well. Use of envelope-packed films avoids many of these problems until the envelope is opened for processing.

Another important precaution is to avoid drawing film rapidly from cartons, exposure holders or cassettes. Such care will help to eliminate circular or treelike black markings in the radiograph that sometimes result due to static electric discharges.

■ 〖Point 3〗 Exposure Vaults & Cabinets

Exposure vaults and cabinets allow personnel to work safely in the area while exposures are taking place. Exposure vaults tend to be larger walk in rooms with shielding provided by high-density concrete block and lead.

Exposure cabinets (Fig. 5–26) are often self-contained units with integrated X-ray equipment and are typically shielded with steel and lead to absorb X-ray radiation.

Exposure vaults and cabinets are equipped with protective interlocks that disable the system if anything interrupts the integrity of the enclosure. Additionally, walk in vaults are equipped with emergency "kill buttons" that allow radiographers to shut down the system if it should accidentally be started while they were in the vault.

Fig. 5–26　Exposure Cabinets

Put into Practice

1. Translate the content in Fig. 5–27.

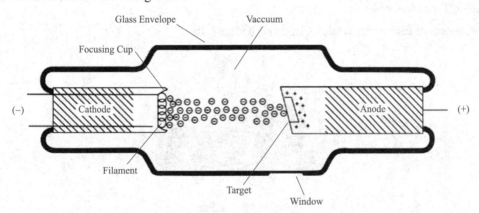

Fig. 5–27　X Tube

2. Read the introduction to the structurix X-ray film (Fig. 5–28) to explain its characteristics.

Structurix X-ray Film

High Quality Radiographic Film.

The Structurix radiographic film product family builds on two critical facets of advanced film technology: high quality images, and rugged performance. Structurix' reputation for excellence is the result of Waygate Technologies' continuous striving for the highest product quality. We offer a full range of Structurix radiographic films from ultrafine grain to high speed, in all standard sheet sizes and on rolls. For all of your requirement needs, we have a structurix film with the right characteristics and packaging.

Optimum Image Quality

• Featuring an AGFA emulsion breakthrough that provides increased contrast and maximum detail perceptibility, resulting in the highest intrinsic defect recognition for each speed range

- Interpret even the smallest details with ease
- High quality finished X-ray film with a brilliant surface and pleasant blue tint

Protective Coating

- Special protective top coating, resulting directly from the Split Antistress Layer (SAL) technology
- Unique high resistance to pressure, scratching and creasing

Consistent Production Quality

- Films meet the most rigorous worldwide quality standards
- Produced at a single facility under tightly controlled conditions in an ultramodern coating room
- Utilizes a system built using Total Quality Management approach, certified by the ISO 9001-2000 label, providing exceptional performance in production consistency

Consistent Processing Quality

- Uses Cubic Grain Plus technology
- Providing consistent and excellent results over a wide range of operating conditions Multiple Packaging Types Available
- Economical darkroom packaging in sheets and on rolls

Fig. 5-28 X Film

5.5 Techniques & Calibrations

〖Point 1〗 Image Considerations

The usual objective in radiography is to produce an image showing the highest amount of detail possible. This requires careful control of a number of different variables that can affect image quality. Radiographic sensitivity is a measure of the quality of an image in terms of the smallest detail or discontinuity that may be detected. Radiographic sensitivity is dependant on the combined effects of two independent sets of variables. One set of variables affects the contrast and the other set of variables affects the definition of the image (Fig. 5-29).

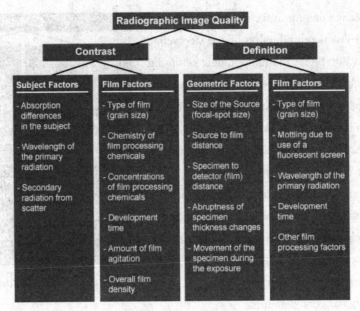

Fig. 5-29 RT Quality

Radiographic contrast (Fig. 5-30) is the degree of density difference between two areas on a radiograph. Contrast makes it easier to distinguish features of interest, such as defects, from the surrounding area. The image to the right shows two radiographs of the same stepwedge. The upper radiograph has a high level of contrast and the lower radiograph has a lower level of contrast. While they are both imaging the same change in thickness, the high contrast image uses a larger change in radiographic density to show this change. In each of the two radiographs,

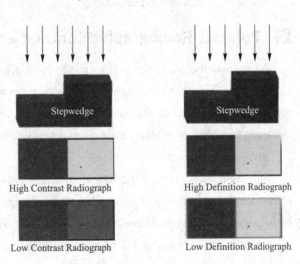

Fig. 5-30 Radiographic Contrast

there is a small circle, which is of equal density in both radiographs. It is much easier to see in the high contrast radiograph. The factors affecting contrast will be discussed in more detail on the following page.

Radiographic definition is the abruptness of change in going from one area of a given radiographic density to another. Like contrast, definition also makes it easier to see features of interest, such as defects, but in a totally different way. In the image to the right, the upper radiograph has a high level of definition and the lower radiograph has a lower level of definition. In the high definition radiograph it can be seen that a change in the thickness of the stepwedge translates to an abrupt change in radiographic density. It can be seen that the details, particularly the small circle, are much easier to see in the high definition radiograph. It can be said that the detail portrayed in the radiograph is equivalent to the physical change present in the stepwedge. In other words, a faithful visual reproduction of the stepwedge was produced. In Fig. 5-31, the radiographic setup did not produce a faithful visual reproduction. The edge line between the steps is blurred. This is evidenced by the gradual transition between the high and low density areas on the radiograph. The factors affecting definition will be discussed in more detail on a following page.

Since radiographic contrast and definition are not dependent upon the same set of factors, it is possible to produce radiographs with the following qualities:

- Low contrast and poor definition
- High contrast and poor definition
- Low contrast and good definition
- High contrast and good definition

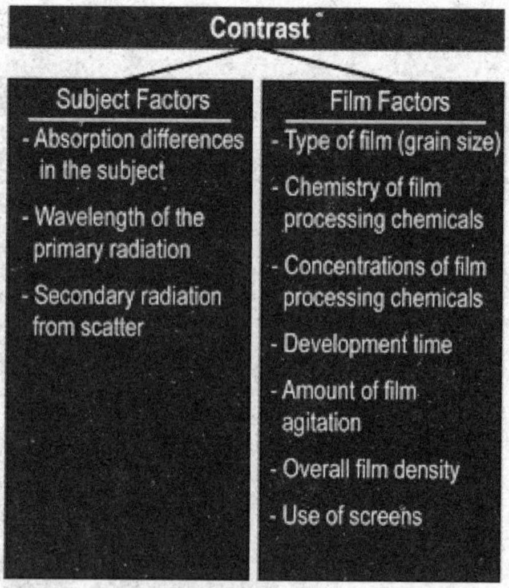

Fig. 5-31 Contrast Factors

〖Point 2〗 Radiographic Contrast

As mentioned on the previous page, radiographic contrast describes the differences in photographic density in a radiograph. The contrast between different parts of the image is what forms the image and the greater the contrast, the more visible features become. Radiographic contrast has two main contributors: subject contrast and detector (film) contrast.

Subject Contrast

Subject contrast is the ratio of radiation intensities transmitted through different areas of the component being evaluated. It is dependant on the absorption differences in the component, the wavelength of the primary radiation, and intensity and distribution of secondary radiation due to scattering.

It should be no surprise that absorption differences within the subject will affect the level of contrast in a radiograph. The larger the difference in thickness or density between two areas of the subject, the larger the difference in radiographic density or contrast. However, it is also possible to radiograph a particular subject and produce two radiographs having entirely different contrast levels. Generating X-rays using a low kilovoltage will generally result in a radiograph with high contrast. This occurs because low energy radiation is more easily attenuated. Therefore, the ratio of photons that are transmitted through a thick and thin area will be greater with low energy radiation. This in turn will result in the film being exposed to a greater and lesser degree in the two areas (Fig. 5-32).

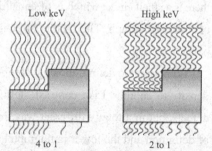

Fig. 5-32 Result of Diffrent Kilovoltage

There is a tradeoff, however. Generally, as contrast sensitivity increases, the latitude of the radiograph decreases. Radiographic latitude refers to the range of material thickness that can be imaged. This means that more areas of different thicknesses will be visible in the image. Therefore, the goal is to balance radiographic contrast and latitude so that there is enough contrast to identify the features of interest but also to make sure the latitude is great enough so that all areas of interest can be inspected with one radiograph. In thick parts with a large range of thicknesses, multiple radiographs will likely be necessary to get the necessary density levels in all areas.

Film Contrast

Film contrast refers to density differences that result due to the type of film used, how it was exposed, and how it was processed. Since there are other detectors besides film, this could be called detector contrast, but the focus here will be on film. Exposing a film to produce higher film densities will generally increase the contrast in the radiograph.

A typical film characteristic curve, which shows how a film responds to different amounts of radiation exposure, is shown to Table 5–4. (More information on film characteristic curves is presented later in this section.) From the shape of the curves, it can be seen that when the film has not seen many photon interactions (which will result in a low film density) the slope of the curve is low. In this region of the curve, it takes a large change in exposure to produce a small change in film density. Therefore, the sensitivity of the film is relatively low. It can be seen that changing the log of the relative exposure from 0.75 to 1.4 only changes the film density from 0.20 to about 0.30. However, at film densities above 2.0, the slope of the characteristic curve for most films is at its maximum. In this region of the curve, a relatively small change in exposure will result in a relatively large change in film density. For example, changing the log of relative exposure from 2.4 to 2.6 would change the film density from 1.75 to 2.75. Therefore, the sensitivity of the film is high in this region of the curve. In general, the highest overall film density that can be conveniently viewed or digitized will have the highest level of contrast and contain the most useful information.

Table 5–4 Typical Film Characteristic Curve

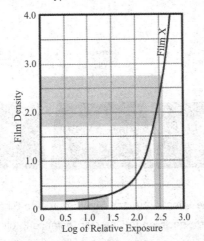

Lead screens in the thickness range of 0.004 to 0.015 inch typically reduce scatter radiation at energy levels below 150 000 Volts. Above this point they will emit electrons to provide more exposure of the film to ionizing radiation, thus increasing the density and contrast of the radiograph. Fluorescent screens produce visible light when exposed to radiation and this light further exposes the film and increases contrast.

〖Point 3〗 Definition

As mentioned previously, radiographic definition is the abruptness of change from one density to another. Geometric factors of the equipment and the radiographic setup, and film and screen factors both have an effect on definition. Geometric factors include the size of the area of origin of the radiation, the source-to-detector (film) distance, the specimen-to-detector (film) distance, movement of the source, specimen or detector during exposure, the angle between the source and some feature and the abruptness of change in specimen thickness or density (Fig. 5-33).

Geometric Factors

The effect of source size, source-to-film distance and the specimen-to-detector distance were covered in detail on the geometric unsharpness page. But briefly, to produce the highest level of definition, the focal-spot or source size should be as close to a point source as possible, the source-to-detector distance should be a great as practical, and the specimen-to-detector distance should be a small as practical. This is shown graphically in (Fig. 5-34).

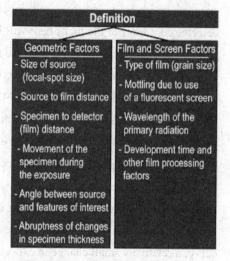

Fig. 5-33 Geometric Factors and Film and Screen Factors

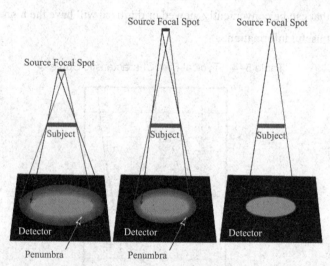

Fig. 5-34 The Effect of Source Size, Source-to-Film Distance and the Specimen-to-Detector Distance Factors

The angle between the radiation and some features will also have an effect on definition (Fig. 5-35). If the radiation is parallel to an edge or linear discontinuity, a sharp distinct boundary will be seen in the image. However, if the radiation is not parallel with the discontinuity, the feature will appear distorted, out of position and less defined in the image. Abrupt changes in thickness and/or density will appear more defined in a radiograph than will areas of gradual change. For example, consider a circle. Its largest dimension will a cord that passes through its centerline. As the cord is moved away from the centerline, the thickness gradually decreases. It is sometimes difficult to locate the edge of a void due to this gradual change in thickness.

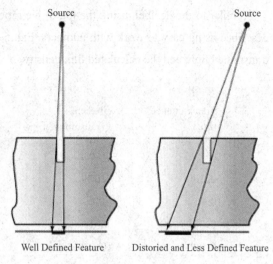

Fig. 5–35　Angle Factors

Lastly, any movement of the specimen, source or detector during the exposure will reduce definition. Similar to photography, any movement will result in blurring of the image. Vibration from nearby equipment may be an issue in some inspection situations.

Film and Screen Factors

The last set of factors concern the film and the use of fluorescent screens. A fine grain film is capable of producing an image with a higher level of definition than is a coarse grain film. Wavelength of the radiation will influence apparent graininess. As the wavelength shortens and penetration increases, the apparent graininess of the film will increase. Also, increased development of the film will increase the apparent graininess of the radiograph.

The use of fluorescent screens also results in lower definition. This occurs for a couple of different reasons. The reason that fluorescent screens are sometimes used is because incident radiation causes them to give off light that helps to expose the film. However, the light they produce spreads in all directions, exposing the film in adjacent areas, as well as in the areas which are in direct contact with the incident radiation. Fluorescent screens also produce screen mottle on radiographs. Screen mottle is associated with the statistical variation in the numbers of photons that interact with the screen from one area to the next.

〖Point 4〗 Radiographic Density

Radiographic density (AKA optical, photographic or film density) is a measure of the degree of film darkening. Technically it should be called "transmitted density" when associated with transparent-base film since it is a measure of the light transmitted through the film. Radiographic density is the logarithm of two measurements: the intensity of light incident on the film (I_0) and the intensity of light transmitted through the film (I_t). This ratio is the inverse of transmittance.

$$D = \log \frac{I_0}{I_t}$$

Similar to the decibel, using the log of the ratio allows ratios of significantly different sizes to be described using easy to work with numbers. Fig. 5-36 shows the relationship between the amount of transmitted light and the calculated film density.

Transmittance (I_t/I_0)	Percent Transmittance	Inverse of Transmittance (I_0/I_t)	Film Density [Log (I_0/I_t)]
0	100%	1	0
0.1	10%	10	1
0.01	1%	100	2
0.001	0.1%	1000	3
0.0001	0.01%	10 000	4
0.00001	0.001%	100 000	5
0.000001	0.0001%	1 000 000	6
0.0000001	0.00001%	10 000 000	7

Fig. 5-36 Ratio

From this table, it can be seen that a density reading of 2.0 is the result of only one percent of the incident light making it through the film. At a density of 4.0 only 0.01% of transmitted light reaches the far side of the film. Industrial codes and standards typically require a radiograph to have a density between 2.0 and 4.0 for acceptable viewing with common film viewers. Above 4.0, extremely bright viewing lights is necessary for evaluation. Contrast within a film increases with increasing density, so in general, the higher the density the better. When radiographs will be digitized, densities above 4.0 are often used since digitization systems can capture and redisplay for easy viewing information from densities up to 6.0.

〖Point 5〗 Film Characteristic Curves

In film radiography, the number of photons reaching the film determines how dense the film will become when other factors such as the developing time are held constant. The number of photons reaching the film is a function of the intensity of the radiation and the time that the film is exposed to the radiation. The term used to describe the control of the number of photons reaching the film is "exposure".

Film Characteristic Curves

Different types of radiographic film respond differently to a given amount of exposure. Film manufacturers commonly characterize their film to determine the relationship between the applied exposure and the resulting film density (Fig. 5-37). This relationship commonly varies over a range of film densities, so the data is presented in the form of a curve such as the one for Kodak AA400 shown to the right. The plot is called a film characteristic curve, sensitometric curve, density curve, or H and D curve (named for developers Hurter and Driffield). "Sensitometry" is the science of measuring the response of photographic emulsions to light or radiation.

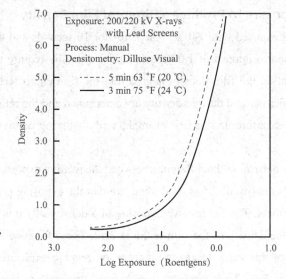

Fig. 5–37 Relationship Between the Applied Exposure and the Resulting Film Density

A log scale is used or the values are reported in log units on a linear scale to compress the x-axis. Also, relative exposure values (unitless) are often used. Relative exposure is the ratio of two exposures. For example, if one film is exposed at 100 keV for 6 mA·min and a second film is exposed at the same energy for 3 mA·min, then the relative exposure would be 2. The image directly to the right shows three film characteristic curves with the relative exposure plotted on a log scale, while Fig. 5–38 shows the log relative exposure plotted on a linear scale.

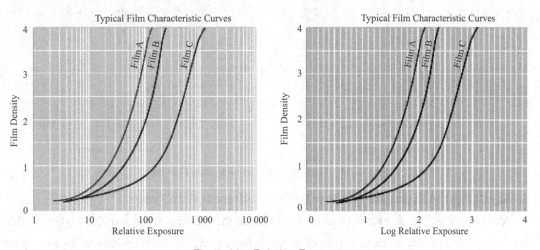

Fig. 5–38 Relative Exposure

Use of the logarithm of the relative exposure scale makes it easy to compare two sets of values, which is the primary use of the curves. Film characteristic curves can be used to adjust the exposure used to produce a radiograph with a certain density to an exposure that will produce a second radiograph of higher or lower film density. The curves can also be used to relate the exposure produced with one type of film to exposure needed to produce a radiograph of the same density with a second type of film.

Adjusting the Exposure to Produce a Different Film Density

Suppose Film B was exposed with 140 keV at 1 mA for 10 seconds and the resulting radiograph had a density in the region of interest of 1.0. Specifications typically require the density to be above 2.0 for reasons discussed on the film density page. From the film characteristic curve, the relative exposures for the actual density and desired density are determined and the ratio of these two values is used to adjust the actual exposure. In this first example, a plot with log relative exposure and a linear x-axis will be used.

From Fig. 5–39, first determine the difference between the relative exposures of the actual and the desired densities. A target density of 2.5 is used to ensure that the exposure produces a density above the 2.0 minimum requirement. The log relative exposure of a density of 1.0 is 1.62 and the log of the relative exposure when the density of the film is 2.5 is 2.12. The difference between the two values is 0.5. Take the antilog of this value to change it from log relative exposure to simply the relative exposure and this value is 3.16. Therefore, the exposure used to produce the initial radiograph with a 1.0 density needs to be multiplied by 3.16 to produce a radiograph with the desired density of 2.5. The exposure of the original X-ray was 10 mA·s, so the new exposure must be 10 mA·s × 3.16 or 31.6 mA·s at 140 keV.

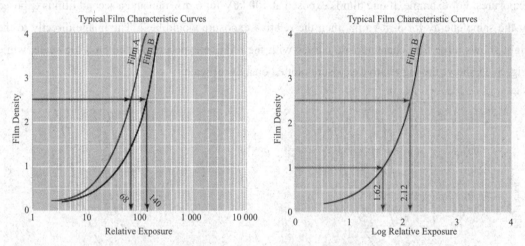

Fig. 5–39 Typical Film Characteristic Curves

Adjusting the Exposure to Allow Use of a Different Film Type

Another use of film characteristic curves is to adjust the exposure when switching types of film. The location of the characteristic curves of different films along the x-axis relates to the film speed of the films. The farther to the right that a curve is on the chart, the slower the film speed. It must be noted that the two curves being used must have been produced with the same radiation energy. The shape of the characteristic curve is largely independent of the wavelength of the X-ray or gamma radiation, but the location of the curve along the x-axis, with respect to the curve of another film, does depend on radiation quality.

Suppose an acceptable radiograph with a density of 2.5 was produced by exposing Film A for 30 seconds at 1 mA and 130 keV. Now, it is necessary to inspect the part using Film B. The exposure can be adjusted by following the above method, as long at the two film characteristic curves were produced with roughly the same radiation quality. For this example, the characteristic curves for Film A and B are shown on a chart showing relative exposure on a log scale. The relative exposure that produced a density of 2.5 on Film A is found to be 68. The relative exposure that should produce a density of 2.5 on Film B is found to be 140. The relative exposure of Film B is about twice that of Film A, or 2.1 to be more exact. Therefore, to produce a 2.5 density radiograph with Film B the exposure should be 30 mA·s times 2.1 or 62 mA·s.

〖Point 6〗 Exposure Calculations

Properly exposing a radiograph is often a trial and error process, as there are many variables that affect the final radiograph. Some of the variables that affect the density of the radiograph include:
- The spectrum of radiation produced by the X-ray generator.
- The voltage potential used to generate the X-rays (keV).
- The amperage used to generate the X-rays (mA).
- The exposure time.
- The distance between the radiation source and the film.
- The material of the component being radiographed.
- The thickness of the material that the radiation must travel through.
- The amount of scattered radiation reaching the film.
- The film being used.
- The concentration of the film processing chemicals and the contact time.

The current industrial practice is to develop a procedure that produces an acceptable density by trail for each specific X-ray generator. This process may begin using published exposure charts to determine a starting exposure, which usually requires some refinement.

However, it is possible to calculate the density of a radiograph to a fair degree accuracy when the spectrum of an X-ray generator has been characterized. The calculation cannot completely account for scattering but, otherwise, the relationship between many of the variables and their effect on film density is known. Therefore, the change in film density can be estimated for any given variable change. For example, from Newton's Inverse Square Law, it is known that the intensity of the radiation varies inversely with distance from the source. It is also known that the intensity of the radiation transmitted through a material varies exponentially with the linear attenuation coefficient (m) and the thickness of the material.

A number of radiographic modeling program are available that make this calculation. These programs can provide a fair representation of the radiograph that will be produce with a specific setup and parameters. The applet below is a very simple radiographic density calculator. The applet allows the density of a radiograph to be estimated based on material, thickness, geometry, energy (voltage), current, and time. The effect of the energy and the physical setup are shown by looking at the film density after exposure. Since the calculation uses a generic (and fixed characteristic) X-ray source, fixed film type and development, the applet results will differ considerably from industrial X-ray configurations. The applet is design simply to demonstrate the affects of the variable on the resulting film density.

How to Use This Applet?

First choose a material. Each material has a mass attenuation constant, *mu*. Next, the voltage to the X-ray source needs to be set. Continue to fill in numbers for the rest of the variables. The current is the number of milliamps that flow to the source. After the Distance, Time and Thickness have been set, press the "Calculate" button.

Note, the I_0 field has a number in it. This is the initial intensity of the X-ray beam. For large numbers, it may be necessary to use the mouse to see the entire number. Click on the number and move the mouse as if selecting it. The cyan pointer indicates the density on the resultant radiograph. The two other pointers represent under- and over-exposure by a factor of four. These may be used to judge the degree of contrast in the resultant radiograph.

Try the following examples: material: aluminum, kV: 120, mA: 5, distance: 0.5 meter, time: 90 seconds, thickness: 6.5 cm. The resultant density will be 2.959. As can be noted on the stepwedge, reducing the exposure by a factor of four will change the density to a value of 1.0, and increasing the exposure by a factor of four will result in a density of 5.0. Reduce the time from 90 seconds to 22.5 seconds (factor of four) and note the results.

Change the material to iron and press "Calculate". Note that not enough radiation is received to generate an image. Change the following: kV, 320; mA, 10; time, 900 seconds; thickness, 1.25 cm; and then click "Calculate". Note the resulting center density of 0.561. With aluminum, the time was altered by a factor of four to change the density. With the iron, current (mA) must be increased by a factor of four to produce an increase in density. Change the current from 10 to 40 and calculate the results.

Caution: This applet does not have knowledge of the characteristics of any particular real-life X-ray source and should not be used other than as a theoretical tool for making predictions of exposure and contrast.

〖Point 7〗 Controlling Radiographic Quality

One of the methods of controlling the quality of a radiograph is through the use of image quality indicators (IQIs) . IQIs, which are also referred to as penetrameters, provide a means of visually informing the film interpreter of the contrast sensitivity and definition of the radiograph. The IQI indicates that a specified amount of change in material thickness will be detectable in the radiograph, and that the radiograph has a certain level of definition so that the density changes are not lost due to unsharpness. Without such a reference point, consistency and quality could not be maintained and defects could go undetected.

Image quality indicators take many shapes and forms due to the various codes or standards that invoke their use. In the United States, two IQI styles are prevalent: the placard, or hole-type and the wire IQI. IQIs comes in a variety of material types so that one with radiation absorption characteristics similar to the material being radiographed can be used.

Hole-Type IQIs

ASTM Standard E1025 gives detailed requirements for the design and material group classification of hole-type image quality indicators. E1025 designates eight groups of shims based on their radiation absorption characteristics. A notching system is incorporated into the requirements, which allows the radiographer to easily determine if the IQI is the correct material type for the product. The notches in the IQI to the right indicate that it is made of aluminum. The thickness in thousands of an inch is noted on each pentameter by one or more lead number. The IQI to the right (Fig. 5–40) is 0.005 inch thick. IQIs may also be manufactured to a military or other industry specification and the material type and thickness may be indicated differently. For example, the IQI on the left in the image above uses lead letters to indicate the material. The numbers on this same IQI indicate the sample thickness that the IQI would typically be placed on when attempting to achieve two percent contrast sensitivity.

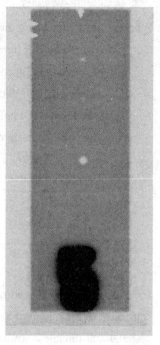

Fig. 5–40 IQI Film Display

Image quality levels are typically designated using a two-part expression such as 2-2T. The first term refers to the IQI thickness expressed as a percentage of the region of interest of the part being inspected. The second term in the expression refers to the diameter of the hole that must be revealed and it is expressed as a multiple of the IQI thickness. Therefore, a 2-2T call-out would mean that the shim thickness should be two percent of the material thickness and that a hole that is twice the IQI thickness must be detectable on the radiograph. This presentation of a 2-2T IQI in the radiograph verifies that the radiographic technique is capable of showing a material loss of 2% in the area of interest.

It should be noted that even if 2-2T sensitivity is indicated on a radiograph, a defect of the same diameter and material loss may not be visible. The holes in the IQI represent sharp boundaries, and a small thickness change. Discontinues within the part may contain gradual changes and are often less visible. The IQI is used to indicate the quality of the radiographic technique and not intended to be used as a measure of the size of a cavity that can be located on the radiograph.

Wire IQIs

ASTM Standard E747 covers the radiographic examination of materials using wire IQIs to control image quality. Wire IQIs consist of a set of six wires arranged in order of increasing diameter and encapsulated between two sheets of clear plastic. E747 specifies four wire IQI sets, which control the wire diameters. The set letter (A, B, C or D) is shown in the lower right corner of the IQI. The number in the lower left corner indicates the material group. The same image quality levels and expressions (i.e. 2-2T) used for hole-type IQIs are typically also used for wire IQIs. The wire sizes that correspond to various hole-type quality levels can be found in a table in E747 or can be calculated using the following formula.

$$F^3 d^3 l = T^2 H^2 \left(\frac{\pi}{4}\right)$$

Where: F—0.79 (constant form factor for wire);

d—wire diameter (mm or inch);

l—7.6 mm or 0.3 inch (effective length of wire);

T—Hole-type IQI thickness (mm or inch);

H—Hole-type IQI hole diameter (mm or inch).

Placement of IQIs is shown in Fig. 5-41.

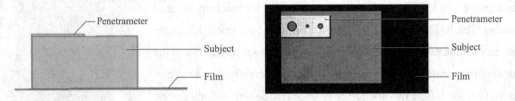

Fig. 5-41 Placement of IQIs

IQIs should be placed on the source side of the part over a section with a material thickness equivalent to the region of interest. If this is not possible, the IQI may be placed on a block of similar material and thickness to the region of interest. When a block is used, the IQI should be the same distance from the film as it would be if placed directly on the part in the region of interest. The IQI should also be placed slightly away from the edge of the part so that at least three of its edges are visible in the radiograph.

〖Point 8〗 Film Processing

As mentioned previously, radiographic film consists of a transparent, blue-tinted base coated on both sides with an emulsion. The emulsion consists of gelatin containing microscopic, radiation sensitive silver halide crystals, such as silver bromide and silver chloride. When X-rays, gamma rays or light rays strike the the crystals or grains, some of the Br^- ions are liberated and captured by the Ag^+ ions. In this condition, the radiograph is said to contain a latent (hidden) image because the change in the grains is virtually undetectable, but the exposed grains are now more sensitive to reaction with the developer.

When the film is processed (Fig. 5-42), it is exposed to several different chemicals solutions for controlled periods of time. Processing film basically involves the following five steps.

• Development—The developing agent gives up electrons to convert the silver halide grains to metallic silver. Grains that have been exposed to the radiation develop more rapidly, but given enough time the developer will convert all the silver ions into silver metal. Proper temperature control is needed to convert exposed grains to pure silver while keeping unexposed grains as silver halide crystals.

- Stopping the development—The stop bath simply stops the development process by diluting and washing the developer away with water.
- Fixing—Unexposed silver halide crystals are removed by the fixing bath. The fixer dissolves only silver halide crystals, leaving the silver metal behind.
- Washing—The film is washed with water to remove all the processing chemicals.
- Drying—The film is dried for viewing.

Processing film is a strict science governed by rigid rules of chemical concentration, temperature, time, and physical movement. Whether processing is done by hand or automatically by machine, excellent radiographs require a high degree of consistency and quality control.

Fig. 5–42 Physical Film Display

Manual Processing & Darkrooms

Manual processing begins with the darkroom. The darkroom should be located in a central location, adjacent to the reading room and a reasonable distance from the exposure area. For portability, darkrooms are often mounted on pickups or trailers.

Film should be located in a light, tight compartment, which is most often a metal bin that is used to store and protect the film. An area next to the film bin that is dry and free of dust and dirt should be used to load and unload the film. Another area, the wet side, should be used to process the film. This method protects the film from any water or chemicals that may be located on the surface of the wet side.

Each of step in the film processing must be excited properly to develop the image, wash out residual processing chemicals, and to provide adequate shelf life of the radiograph. The objective of processing is two fold: first, to produce a radiograph adequate for viewing; and second, to prepare the radiograph for archival storage. Radiographs are often stored for 20 years or more as a record of the inspection.

Automatic Processor Evaluation

The automatic processor is the essential piece of equipment in every X-ray department. The automatic processor will reduce film processing time when compared to manual development by a factor of four. To monitor the performance of a processor, apart from optimum temperature and mechanical checks, chemical and sensitometric checks should be performed for developer and fixer. Chemical checks involve measuring the pH values of the developer and fixer as well as both replenishers. Also, the specific gravity and fixer silver levels must be measured. Ideally, pH should be measured daily and it is important to record these measurements, as regular logging provides very useful information. The daily measurements of pH values for the developer and fixer can then be plotted to observe the trend of variations in these values compared to the normal pH operating levels to identify problems.

Sensitometric checks may be carried out to evaluate if the performance of films in the automatic processors is being maximized. These checks involve measurement of basic fog level, speed and average gradient made at 1 ℃ intervals of temperature. The range of temperature measurement depends on the type of chemistry in use, whether cold or hot developer. These three measurements: fog level, speed and average gradient, should then be plotted against temperature and compared with the manufacturer's supplied figures.

■ 〖Point 9〗 Viewing Radiographs (Fig. 5–43)

Radiographs (developed film exposed to X-ray or gamma radiation) are generally viewed on a light-box. However, it is becoming increasingly common to digitize radiographs and view them on a high resolution monitor. Proper viewing conditions are very important when interpreting a radiograph. The viewing conditions can enhance or degrade the subtle details of radiographs.

Viewing Radiographs

Before beginning the evaluation of a radiograph, the viewing equipment and area should be considered. The area should be clean and free of distracting materials. Magnifying aids, masking aids, and film markers should

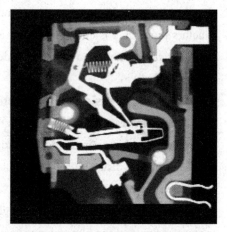

Fig. 5–43 Electrical Circuit Breaker

be close at hand. Thin cotton gloves should be available and worn to prevent fingerprints on the radiograph. Ambient light levels should be low. Ambient light levels of less than 2 fc are often recommended, but subdued lighting (rather than total darkness) is preferable in the viewing room. The brightness of the surroundings should be about the same as the area of interest in the radiograph. Room illumination must be arranged so that there are no reflections from the surface of the film under examination.

Film viewers should be clean and in good working condition. There are four groups of film viewers. These include strip viewers, area viewers, spot viewers, and a combination of spot and area viewers. Film viewers should provide a source of defused, adjustable, and relativity cool light as heat from viewers can cause distortion of the radiograph. A film having a measured density of 2.0 will allow only 1% of the incident light to pass. A film containing a density of 4.0 will allow only 0.01% of the incident light to pass. With such low levels of light passing through the radiograph, the delivery of a good light source is important.

The radiographic process should be performed in accordance with a written procedure or code, or as required by contractual documents. The required documents should be available in the viewing area and referenced as necessary when evaluating components. Radiographic film quality and acceptability, as required by the procedure, should first be determined. It should be verified that the radiograph was produced to the correct density on the required film type, and that it contains the correct identification information. It should also be verified that the proper image quality indicator was used and that the required sensitivity level was met. Next, the radiograph should be checked to ensure that it does not contain processing and handling artifacts that could mask discontinuities or other details of interest. The technician should develop a standard process for evaluating the radiographs so that details are not overlooked.

Once a radiograph passes these initial checks, it is ready for interpretation. Radiographic film interpretation is an acquired skill combining visual acuity with knowledge of materials, manufacturing processes, and their associated discontinuities. If the component is inspected while in service, an understanding of applied loads and history of the component is helpful. A process for viewing radiographs (e.g. left to right, top to bottom, etc.) is helpful and will prevent overlooking an area on the radiograph. This process is often developed over time and individualized. One part of the interpretation process, sometimes overlooked, is rest. The mind as well as the eyes need to occasionally rest when interpreting radiographs.

When viewing a particular region of interest, techniques such as using a small light source and moving the radiograph over the small light source, or changing the intensity of the light source will help the radiographer identify relevant indications. Magnifying tools should also be used when appropriate to help identify and evaluate indications. Viewing the actual component being inspected is very often helpful in developing an understanding of the details seen in a radiograph.

Interpretation of radiographs is an acquired skill that is perfected over time. By using the proper equipment and developing consistent evaluation processes, the interpreter will increase his or her probability of detecting defects.

〖Point 10〗 Radiograph Interpretation—Welds

In addition to producing high quality radiographs, the radiographer must also be skilled in radiographic interpretation. Interpretation of radiographs takes place in three basic steps: detection, interpretation, and evaluation. All of these steps make use of the radiographer's visual acuity. Visual acuity is the ability to resolve a spatial pattern in an image. The ability of an individual to detect discontinuities in radiography is also affected by the lighting condition in the place of viewing, and the experience level for recognizing various features in the image. The following material was developed to help students develop an understanding of the types of defects found in weldments and how they appear in a radiograph.

Discontinuities

Discontinuities are interruptions in the typical structure of a material. These interruptions may occur in the base metal, weld material or "heat affected" zones. Discontinuities, which do not meet the requirements of the codes or specifications used to invoke and control an inspection, are referred to as defects.

General Welding Discontinuities

The following discontinuities are typical of all types of welding.

Cold lap (Fig. 5–44) is a condition where the weld filler metal does not properly fuse with the base metal or the previous weld pass material (interpass cold lap) . The arc does not melt the base metal sufficiently and causes the slightly molten puddle to flow into the base material without bonding.

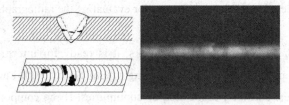

Fig. 5–44 Cold Lap

Porosity (Fig. 5–45) is the result of gas entrapment in the solidifying metal. Porosity can take many shapes on a radiograph but often appears as dark round or irregular spots or specks appearing singularly, in clusters, or in rows. Sometimes, porosity is elongated and may appear to have a tail. This is the result of gas attempting to escape while the metal is still in a liquid state and is called wormhole porosity. All porosity is a void in the material and it will have a higher radiographic density than the surrounding area.

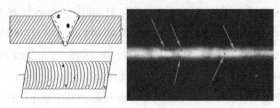

Fig. 5–45 Porosity

Cluster porosity (Fig. 5–46) is caused when flux coated electrodes are contaminated with moisture. The moisture turns into a gas when heated and becomes trapped in the weld during the welding process. Cluster porosity appear just like regular porosity in the radiograph but the indications will be grouped close together.

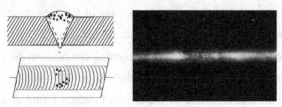

Fig. 5–46 Cluster Porosity

Slag inclusions are nonmetallic solid material entrapped in weld metal or between weld and base metal. In a radiograph, dark, jagged asymmetrical shapes within the weld or along the weld joint areas are indicative of slag inclusions.

Incomplete penetration (IP) or lack of penetration (LOP) occurs when the weld metal fails to penetrate the joint. It is one of the most objectionable weld discontinuities. Lack of penetration allows a natural stress riser from which a crack may propagate. The appearance on a radiograph is a dark area with well-defined, straight edges that follows the land or root face down the center of the weldment.

Incomplete fusion is a condition where the weld filler metal does not properly fuse with the base metal. Appearance on radiograph: usually appears as a dark line or lines oriented in the direction of the weld seam along the weld preparation or joining area.

Internal concavity or suck back (Fig. 5–47) is a condition where the weld metal has contracted as it cools and has been drawn up into the root of the weld. On a radiograph it looks similar to a lack of penetration but the line has irregular edges and it is often quite wide in the center of the weld image.

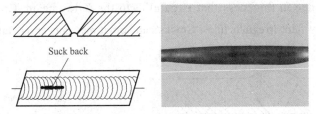

Fig. 5–47 Suck Back

Internal or root undercut is an erosion of the base metal next to the root of the weld. In the radiographic image it appears as a dark irregular line offset from the centerline of the weldment.

Undercutting is not as straight edged as LOP because it does not follow a ground edge.

External or crown undercut (Fig. 5–48) is an erosion of the base metal next to the crown of the weld. In the radiograph, it appears as a dark irregular line along the outside edge of the weld area.

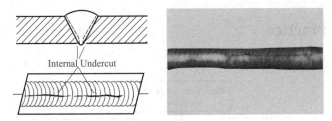

Fig. 5–48 External or Crown Undercut

Offset or mismatch (Fig. 5–49) are terms associated with a condition where two pieces being welded together are not properly aligned. The radiographic image shows a noticeable difference in density between the two pieces. The difference in density is caused by the difference in material thickness. The dark, straight line is caused by the failure of the weld metal to fuse with the land area.

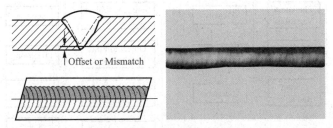

Fig. 5–49 Offset or Mismatch

Inadequate weld reinforcement is an area of a weld where the thickness of weld metal deposited is less than the thickness of the base material. It is very easy to determine by radiograph if the weld has inadequate reinforcement, because the image density in the area of suspected inadequacy will be higher (darker) than the image density of the surrounding base material.

Excess weld reinforcement is an area of a weld that has weld metal added in excess of that specified by engineering drawings and codes. The appearance on a radiograph is a localized, lighter area in the weld. A visual inspection will easily determine if the weld reinforcement is in excess of that specified by the engineering requirements.

Cracks (Fig. 5-50) can be detected in a radiograph only when they are propagating in a direction that produces a change in thickness that is parallel to the X-ray beam. Cracks will appear as jagged and often very faint irregular lines. Cracks can sometimes appear as "tails" on inclusions or porosity.

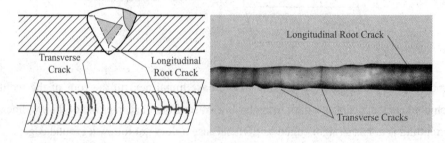

Fig. 5-50   Cracks

Put into Practice

1. Translate Fig. 5-51.

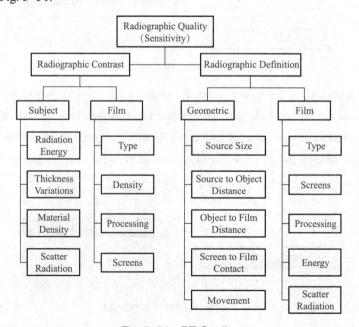

Fig. 5-51   RT Quality

2. Analyze the projection form of Fig. 5-52.

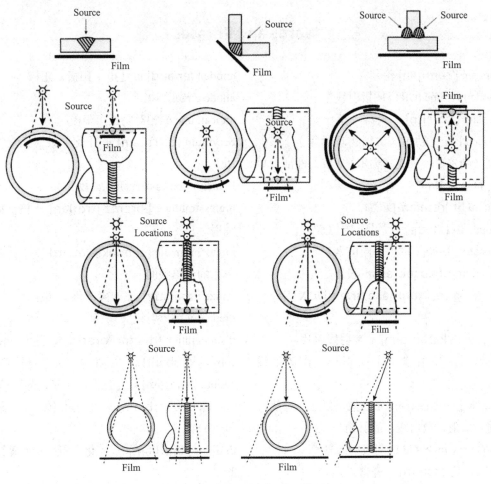

Fig. 5-52 The Projection Form

3. Translate the process card below (Table 5-5).

Table 5-5 RT Report

Work's No. or. Name:			Ref. No:							
Workpiece Name:			Applied Rule:							
NDT Instrument:			Film Type:							
Tube Voltage: kV			Screen							
Exposure: mA·min			Dev. Temp./Time: ℃ /min							
Penetrometer:			Density:							
No.	Test Position	Test No.	Thickness /mm	Defect Kinds						Evaluation Class	Note
				A	B	C	D	E	F		
Note: A-Pore; B-Slag Inclusion; C-Incomplete Penetration; D-Lack of Fusion; E-Crack; F-Undercut											
Acceptance by:			Auditor:				Surveyor:				
Date:			Date:				Date:				

193

Words and Phrases

Roentgen ['rɒntgən] 伦琴
cathode-ray tube (CRT) 阴极射线管
porcelain ['pɔːslɪn] n. 瓷器，瓷
Becquerel 贝克勒尔（法国物理学家）［核］贝克（勒尔），贝克（勒尔）（放射性活度单位，符号为 Bq）
uranium [jʊ'reɪnɪəm] n. 铀
isotope ['aɪsətəʊp] n. ［化］同位素
radioactive isotope 放射性同位素
projection [prə'dʒekʃən] n. 投影
roentgenogram ['rɒntgənəˌgræm] n. X 射线相片
skiagram ['skaɪəgræm] n. X 射线照片
skiagraph n. X 射线照片 vt. 对……作 X 射线摄影
radiation [ˌreɪdɪ'eɪʃən] n. 发散，发光，发热，辐射，放射，放射线，放射物
cathode-ray tube (CRT) 阴极射线管
porcelain ['pɔːsəlɪn] n. 瓷器，瓷
armament ['ɑːməmənt] n. 军备，武器
fusion-welded 熔焊
betatron ['biːtəˌtrɒn] n. 电子感应加速器
American Society of Mechanical Engineers (ASME) 美国机械工程师学会
nonuniform ['nɒn'juːnɪˌfɔːm] adj. 不一致的，不均匀的
radioscopy [ˌreɪdɪ'ɒskəpɪ] n. X 射线透视检验法，射线检查法
real-time radiography 实时射线照相法
Xerography n. 静电复印术
selenium [sɪ'liːnɪəm] n. ［化］硒（元素符号 Se）
positron ['pɒzɪˌtrɒn] n. 正电子
tungsten ['tʌŋstən] n. ［化］钨
impinge [ɪm'pɪndʒ] v. 撞击

anode ['ænəʊd] n. ［电］阳极，正极
anode target 阳极
anticathode n. 相对阴极，阳极
beryllium [bə'rɪlɪəm] n. ［化］铍（元素符号 Be）
Planck's constant 普朗克常数
bremsstrahlung ['bremzˌʃtrɑːləŋ] ［核］韧致辐射
monochromatic [ˌmɒnəkrə'mætɪk] adj. ［物］单色的，单频的
ionize ['aɪənaɪz] vt. 使离子化 vi. 电离
grenz ray 跨界射线
disintegration [dɪsˌɪntɪ'greɪʃn] n. 裂变，衰变
cobalt ['kəʊbɔːlt] n. ［化］钴（元素符号 Co）
cesium ['siːzɪəm] n. ［化］铯（元素符号 Ce）
ytterbium [ɪ'tɜːbɪəm] n. ［化］镱（元素符号 Yb）
thulium ['θjuːlɪəm] n. ［化］铥（元素符号 Tm）
half-life 半衰期
decay constant 衰变常数
specificactivity 比活度
Avogadro's number 阿伏伽德罗常数
Lambert ['læmbət] n. ［姓］兰伯特［物］朗伯（亮度单位）
diverge [daɪ'vɜːdʒ] v. 分散
Beer's Law 比尔定律
Thomson scattering 汤姆森散射
photoelectric effect 光电效应
Compton scattering 康普顿散射
Pair production 电子对
triplet ['trɪplət] n. 三个一组，三份
incoherent [ˌɪnkəʊ'hɪərənt] adj. 不相干的
photodisintegration [ˌfəʊtəʊdɪsˌɪntə'greɪʃn]

n. [核]光致裂变
photon ['fəʊtɒn] *n.* [物]光子
auger ['ɔːgə] *n.* 螺旋
gelatin ['dʒelətɪn] *n.* 凝胶，白明胶
matrix ['meɪtrɪks] *n.* 基体，基质，矩阵
exposure [ɪk'spəʊzə] *n.* 曝光量
exposure time 曝光时间
contrast ['kɒntræst] *n.* 对比度（性、率、法），反差，衬度（比）
emulsion [ɪ'mʌlʃn] *n.* [摄]感光乳剂，乳状液
fluorescent screen 荧光屏
dimensionless [dai'menʃənlis] *adj.* 无量纲的
latitude in exposure [摄]曝光宽容度
film latitude 胶片宽容度
film speed 胶片感光度
fog density 灰雾度
base density 片基光学密度（片基黑度）
radioscopic [ˌreɪdɪəʊ'skɒpɪk] *adj.* 放射镜的，放射观察法的
fluoroscopic [ˌflʊərəskɒpɪk] *adj.* 荧光镜的，荧光检查法的
photocathode [fəʊtəʊ'kæθəʊd] *n.* [电子]光电阴极
line pair 线对
unsharpness *n.* 不清晰度
penumbra [pə'nʌmbrə] *n.* 半影
image quality indicator (IQI) 像质计
resolution [ˌrezə'luːʃn] *n.* 分辨率
resolving power *n.* (光学仪器等)分辨能力
geometric unsharpness 几何不清晰度
motion unsharpness 运动不清晰度
inherent film unsharpness 胶片固有不清晰度
megavoltage *n.* [电]兆伏数，兆伏级
thermal conductivity 热导率，导热性
annular ring emitting 周向辐射
kinetic energy 动能
relativistic [ˌrelətɪ'vɪstɪk] *adj.* 相对论的
attenuation coefficient 衰减系数
absorption coefficient 吸收系数
half-value Thickness (HVT) or half-value layer (HVL) 半价层，半值层
broad beam 宽束
narrow beam 窄束
mean free path 平均自由行程
absorption edge 吸收限，吸收边界
atomic number 原子序数
amorphous [ə'mɔːfəs] *adj.* 无定形的，非晶的
weighted average 加权平均值
mass absorption coefficient 质量吸收系数
gram-atomic absorption coefficient 克原子吸收系数
atomic absorption coefficien 原子吸收系数
versus ['vɜːsəs] *prep.* 与……相对
ordinate ['ɔːdɪnət] *n.* [数]纵线，纵坐标
exponential [ˌekspə'nenʃl] *n.* 指数 *adj.* 指数的
focal spot 焦点
concentric [kən'sentrɪk] *adj.* 同中心的
tenth-value layers (TVL) 十分之一价层
ionization chamber 电离箱，电离室
Geiger counter [核]盖革(-缪勒)计数器
scintillation counting 闪烁计数，用闪烁法测量放射性强度
calorimetry [ˌkælə'rɪmɪtri] *n.* 热量测定法
semi-conductor detectors 半导体探测器
thermoluminescence *n.* 热致发光
intensifying screen 增感屏，光增强屏
inorganic [ˌɪnɔː'gænɪk] *adj.* 无生物的，无机的
image intensifier 图像增强器，图像增强管
halide ['hælaɪd] *n.* [化]卤化物 *adj.* 卤化物的
latent image 潜影
bromide ['brəʊmaɪd] *n.* [化]溴化物
silver bromide 溴化银

Magnetic Particle Inspection Procedure

1. Scope

(1) This procedure describes the general requirements for radiography examination (RT) according to related approved weld map for the metallic welding and casting as may be required by the specification or under which component is being designed and manufactured.

(2) This radiographic testing procedure provides the material, equipment, calibration, personnel qualification, examination process, evaluation, records and acceptance standards for XXX Project which will be fabricated in YYY.

2. Sureace condition

According to T.222.2, the weld ripples or weld surface irregularities on the both the inside (where accessible) and out side shall be removed by any suitable process to such a degree that the resulting radiographic testing image due to any surface irregularities cannot mask or be confused with the image of any discontinuity. The finished surface of all butt welded joints mat be flush with the base material or may have reasonably uniform crowns, with reinforcement not to exceed that specified in the referencing code section.

3. Radiation source

(1) X-radiation:

The radiography testing techniques shall demonstrate that the required radiography sensitivity has been obtained. Maximum X-ray voltage is 300 kV.

(2) Gamma radiation:

The recommended minimum thickness for which Radio-active isotopes may be used as Table 5-6:

Table 5-6 Minimum Thickness

Material	Iridium 192	Cobalt 60
Steel	0.75 in	1.50 in
Copper or high nickel copper	0.65 in	1.30 in
Aluminum	2.50 in	—

The maximum thickness for the use of radioactive isotopes is primarily dictated by exposure time, therefore, upper limits are not shown. The minimum recommended thickness limitation may be reduced when the radiography techniques are used to demonstrate that the required radiographic testing sensitivity have been obtained, by purchaser approval.

4. Radigraphic films

Any commercially available industrial radiography films may be used in accordance with SE 1815 (ASTM) standard test method for film system in industrial radiography. Radiography film shall be fine grain high definition, high contrast film (Kodak type AA 400, FUJI 100 or AGFA D7).

5. Screens

Any commercially available intensifying screen, except those of the fluorescent type, may be used. Intensifying screen for X-ray or Gammar ray method divided in two categories: front screen, back screen. Commonly lead screens use with 27 micron thickness. (Front screen)

6. Penetrameter (IQI)

Penetrameters shall be either the whole type or the wire type and shall be manufactured and identified in accordance with the requirements or alternatives allowed in SE 142 or SE 1025 (for whole type) and SE-747 (for wire type), and appending. ASME V 2007 ED & ASME Sec Ⅷ Div Ⅰ ED 2007.

Penetrameters shall consist of those in Table 5-7 for wire type and those in Table 5-8 for hole type. (Wire type IQI shall be used for welds.)

Table 5-7 Wire IQI Designation, Wire Diameter and Wire Identity

Set A			Set B		
Wire Diameter, in.	mm	Wire Identity	Wire Diameter, in.	mm	Wire Idenilty
0.003 2	0.08	1	0.010	0.25	6
0.004	0.10	2	0.013	0.33	7
0.005	0.13	3	0.016	0.41	8
0.006 3	0.16	4	0.020	0.51	9
0.008	0.20	5	0.025	0.64	10
0.010	0.25	6	0.032	0.81	11

Set C			Set D		
Wire Diameter, in.	mm	Wire Identity	Wire Diameter, in.	mm	Wire Identity
0.032	0.81	11	0.100	2.54	16
0.040	1.02	12	0.126	3.20	17
0.050	1.27	13	0.160	4.06	18
0.063	1.60	14	0.200	5.08	19
0.080	2.03	15	0.250	6.35	20
0.100	2.54	16	0.320	8.13	21

Table 5-8 Hole-type IQI Designation, Thickness and Hole Diameters

IQI Designation	IQI Thickness, in. (mm)	1T Hole Diameter, in. (mm)	2T Hole Diameter, in. (mm)	4T Hole Diameter, in. (mm)
5	0.005 (0.13)	0.010 (0.25)	0.020 (0.51)	0.040 (1.02)
7	0.007 5 (0.19)	0.010 (0.25)	0.020 (0.51)	0.040 (1.02)
10	0.010 (0.25)	0.010 (0.25)	0.020 (0.51)	0.040 (1.02)
12	0.012 5 (0.32)	0.012 5 (0.32)	0.025 (0.64)	0.050 (1.27)
15	0.015 (0.38)	0.015 (0.38)	0.030 (0.76)	0.060 (1.52)
17	0.017 5 (0.44)	0.017 5 (0.44)	0.035 (0.89)	0.070 (1.78)

continued

IQI Designation	IQI Thickness, in. (mm)	1T Hole Diameter, in. (mm)	2T Hole Diameter, in. (mm)	4T Hole Diameter, in. (mm)
20	0.020 (0.51)	0.020 (0.51)	0.040 (1.02)	0.080 (2.03)
25	0.025 (0.64)	0.025 (0.64)	0.050 (1.27)	0.100 (2.54)
30	0.030 (0.76)	0.030 (0.76)	0.060 (1.52)	0.120 (3.05)
35	0.035 (0.89)	0.035 (0.89)	0.070 (1.78)	0.140 (3.56)
40	0.040 (1.02)	0.040 (1.02)	0.080 (2.03)	0.160 (4.06)
45	0.045 (1.14)	0.045 (1.14)	0.090 (2.29)	0.180 (4.57)
50	0.050 (1.27)	0.050 (1.27)	0.100 (2.54)	0.200 (5.08)
60	0.060 (1.52)	0.060 (1.52)	0.120 (3.05)	0.240 (6.10)
70	0.070 (1.78)	0.070 (1.78)	0.140 (3.56)	0.280 (7.11)
80	0.080 (2.03)	0.080 (2.03)	0.160 (4.06)	0.320 (8.13)
100	0.100 (2.54)	0.100 (2.54)	0.200 (5.08)	0.400 (10.16)
120	0.120 (3.05)	0.120 (3.05)	0.240 (6.10)	0.480 (12.19)
140	0.140 (3.56)	0.140 (3.56)	0.280 (7.11)	0.560 (14.22)
160	0.160 (4.06)	0.160 (4.06)	0.320 (8.13)	0.640 (16.26)
200	0.200 (5.08)	0.200 (5.08)	0.400 (10.16)	
240	0.240 (6.10)	0.240 (6.10)	0.480 (12.19)	
280	0.280 (7.11)	0.280 (7.11)	0.560 (14.22)	

7. Selection of penetrameter (IQI)

(1) Material. IQIs shall be selected from either the same alloy material group or grade as identified in SE-1025, or SE-747, as applicable, or from an alloy material group or grade with less radiation absorption than the material being radiographed.

(2) Size. The designated hole IQI or essential wire listed in Table 5-9 provided an equivalent IQI sensitivity is maintained. See T-283.2. shall be as specified in Table 5-9. A thinner or thicker hole-type IQI may be substituted for any section thickness.

① Welds with Reinforcements. The thickness on which the IQI is based is the nominal single-wall thickness plus the estimated weld reinforcement not to exceed the maximum permitted by the referencing Code Section. Backing rings or strips shall not be considered as part of the thickness in IQI selection. The actual measurement of the weld reinforcement is not required.

② Welds Without Reinforcements. The thickness on which the IQI is based is the nominal single-wall thickness. Backing rings or strips shall not be considered as part of the weld thickness in IQI selection.

(3) Welds joining dissimilar materials or welds with dissimilar filler metal. When the weld metal is of an alloy group or grade that has a radiation attenuation that differs from the base material, the IQI material selection shall be based on the weld metal and be in accordance with T-276.1. When the density limits of T-282.2 cannot be met with one IQI and the exceptional density area is at the interface of the weld metal and the base metal, the material selection for the additional IQIs shall be based on the base material and is in accordance with T-276.1

Table 5-9 IQI Selection

Nominal Single-wall Material Thickness Range		IQI			
		Source Side		Film Side	
in	mm	Hole-Type Designation	Wire-Type Essential Wire	Hole-Type Designation	Wire-Type Essential Wire
Up to 0.25, incl.	Up to 6.4, incl.	12	5	10	4
Over 0.25 through 0.375	Over 6.4 through 9.5	15	6	12	5
Over 0.375 through 0.50	Over 9.5 through 12.7	17	7	15	6
Over 0.50 through 0.75	Over 12.7 through 19.0	20	8	17	7
Over 0.75 through 1.00	Over 19.0 through 25.4	25	9	20	8
Over 1.00 through 1.50	Over 25.4 through 38.1	30	10	25	9
Over 1.50 through 2.00	Over 38.1 through 50.8	35	11	30	10
Over 2.00 through 2.50	Over 50.8 through 63.5	40	12	35	11
Over 2.50 through 4.00	Over 63.5 through 101.6	50	13	40	12
Over 4.00 through 6.00	Over 101.6 through 152.4	60	14	50	13
Over 6.00 through 8.00	Over 152.4 through 203.2	80	16	60	14
Over 8.00 through 10.00	Over 203.2 through 254.0	100	17	80	16
Over 10.00 through 12.00	Over 254.0 through 304.8	120	18	100	17
Over 12.00 through 16.00	Over 304.8 through 406.4	160	20	120	18
Over 16.00 through 20.00	Over 406.4 through 508.0	200	21	160	20

8. Placement of Radiographic Testing Penetrameter (IQI)

(1) Source side penetrameters:

The penetrameters shall be placed on the source side of the part being examined, except for the condition described in chapter 8.2.

(2) Film side penetrameters:

Sensitivity: The sensitivity required using wire type IQI shall be 2%.

Sensitivity: (Diameter of thinnest wire visible on radiograph / Part thickness at IQI location) × 100

Where inaccessibility prevents hand placing the penetrameter (s) on the source side, it shall be placed on the film side in contact with the part being examined (Fig. 5-53). A lead letter "F" shall be placed adjacent to or on the penetrameter (s).

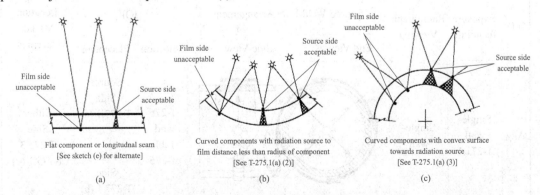

Fig. 5-53 Location Marker Sketches

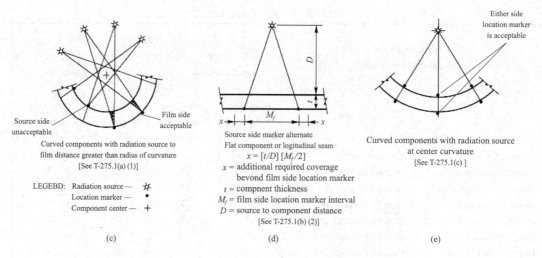

Fig. 5–53 Location Marker Sketches (Cont.)

9. Number of penetrameter (IQI)

When one or more film holders are used for an exposure, ate least one penetrameter imager shall appear on each radiograph.

If the requirements of T-282 are met by using more than one penetrameter, one shall be representative of the lightest area of interest and the other the darkest area of interest.

The intervening densities, on the radiograph, shall be considered as having acceptable density.

Number of IQI shall be according to ASME Sec V. T-277.2.

10. Radiographic testing technique

A single-wall exposure technique shall be used for radiography whenever practical. When it is not practical to use a single-wall radiographic testing technique, a double-wall technique shall be used. An adequate number of exposures shall be made to demonstrate that the required coverage has been obtained.

(1) Single-wall techniques (Table 5–10). In the single-wall radiographic testing technique, the radiation passes through only one wall of the weld (material), which is viewed for acceptance on the radiograph.

Table 5–10 Single-wall Radiographic Techniques

O.D.	Exposure Technique	Radiograph Viewing	Source-Weld-Film Arrangement		IQI		Location Marker Placement
			End View	Side View	Selection	Placement	
Any	Single-wall T-271.1	Single-wall	Exposure Arrangement — A		T-276 and Table T-276	Source Side T-277.1 (a)	Either Side T-275.3 T-275.1 (c)
						Film Side T-277.1 (b)	

O.D.	Exposure Technique	Radiograph Viewing	Source-Weld-Film Arrangement		IQI		Location Marker Placement
			End View	Side View	Selection	Placement	
Any	Single-wall T-271-1	Single-wall	Exposure Arrangement — B		T-276 and Table T-276	Source Side T-277.1 (a)	Film Side T-275.1 (b) (1)
						Film Side T-277.1 (b)	
Any	Single-wall T-271.1	Single-wall	Exposure Arrangement — C		T-276 and Table T-276	Source Side T-277.1 (a)	Source Side T-275.1 (a) (3)
						Film Side T-277.1 (b)	

(2) Double-wall techniques (Table 5–11).

When it is not practical to use a single-wall technique, one of the following double-wall techniques shall be used.

Single-wall viewing. For materials and for welds in components, a technique may be used in which the radiation passes through two walls and only the weld (material) on the film-side wall is viewed for acceptance on the radiograph. When complete coverage is required for circumferential welds (materials), a minimum of three exposures taken 120 deg to each other shall be made.

Double-wall viewing. For materials and for welds in components 3 1/2 in. (89 mm) or less in nominal outside diameter, a technique may be used in which the radiation passes through two walls and the weld (material) in both walls is viewed for acceptance on the same radiograph. For double-wall viewing, only a source-side IQI shall be used. Care should be exercised to ensure that the required geometric unsharpness is not exceeded. If the geometric unsharpness requirement cannot be met, then single-wall viewing shall be used.

For welds, the radiation beam may be offset from the plane of the weld at an angle sufficient to separate the images of the source-side and film-side portions of the weld so that there is no overlap of the areas to be interpreted. When complete coverage is required, a minimum of two exposures taken 90 deg to each other shall be made for each joint.

As an alternative, the weld may be radio graphed with the radiation beam positioned so that the images of both walls are superimposed. When complete coverage is required, a minimum of three exposures taken at either 60 deg or 120 deg to each other shall be made for each joint.

Table 5-11 Double-wall Radiographic Techniques

O.D.	Exposure Technique	Radiograph Viewing	Source-Weld-Film Arrangement		IQI		Location Marker Placement
			End View	Side View	Selection	Placement	
Any	Double-wall: T-271.2 (a) at least 3 Exposures 120 deg to Each Other for Complete Coverage	Single-wall	Exposure Arrangement — D		T-276 and Table T-276	Source Side T-277.1 (a)	Either Side T-275.3 T-275.1 (c)
						Film Side T-277.1 (b)	
Any	Double-wall: T-271.2 (a) at least 3 Exposures 120 deg to Each Other for Complete Coverage	Single-wall	Exposure Arrangement — E		T-276 and Table T-276	Source Side T-277.1 (a)	Film Side T-275.1 (b) (1)
						Film Side T-277.1 (b)	
$3\frac{1}{2}$ in. (88 mm) or less	Double-wall T-271.2 (b) (1) at least 2 Exposures at 90 deg to Each Other for Complete Coverage	Double-wall (Ellipse) : Read Offset Source Side and Film Side Images	Exposure Arrangement — F		T-276 and Table T-276	Source Side T-277.1 (a)	Either Side T-275.2

11. Source to Object and Object to Film Distance (SOD & OFD)

According to geometric unsharpness formula ($U_g = f \times$ OFD/FOD) for minimizing the U_g value, OFD value shall be minimizing therefore object to film distance shall be minimum.

Source to object distance (SOD) shall be set according radiographic technique, object shape and strength of source.

12. Radiographic testing identification system

The method shall be used to produce permanent identification to the radiographies traceable to the contract, components, welds or weld seams, or part numbers, as appropriate. This identification mark shall not obscure the area of interest.

13. Radiographic testing acceptance standard

Refer to ASME Sec Ⅷ, Div. Ⅰ

(1) Butt welded joints surfaces shall be sufficiently free from coarse ripples, grooves, overlaps and abrupt ridges and valleys to permit proper interpretation of radiographic and the required non-destructive examinations.

If there is a question regarding the surface condition of the weld when interpreting a radiographic film, the film shall be compared to the actual weld surface for determination of acceptability.

(2) Indications shown on the radiographies of welds and characterized as imperfections are unacceptable under the following condition:

① Any indications characterized as a crack or zone of incomplete fusion or penetration.

② Any other elongated indication at radiography, which has length greater than:

a. 1/4 in. (6 mm) for t up to 3/4 in. (19 mm)

b. $t/3$ for t from 3/4 in. (19 mm) to 2 1/4 in. (57 mm)

c. $3t/4$ (19 mm) for t over 2 1/4 in. (57 mm)

Where: t— thickness of weld excluding any allowable reinforcement.

③ Any group of aligned indications that have an aggregate length greater than t in a length of 12 t, except when the distance between the successive imperfections exceed $6L$.

Where: L— the length of the longest imperfection in the group.

④ Rounded indications in excess of that specified by the acceptance standards given in ASME Sec. Ⅷ, DIV Ⅰ, appendix 4 Fig. 4-2 to Fig. 4-8.

Note: spot RT shall be done as per ASME Sec. Ⅷ, Div. Ⅰ UW-52; however, the acceptance criteria shall be according to UW-51 (as specification).

14. Defect removal

Repair area shall be located on the weld line after evaluation & interpretation of radiograph. Defects shall be removed by suitable method such as grinding, chipping or gouging (if permitted). welding of the repair area shall meet the requirement of related WPS, PQR.

15. Certification and personnel qualification in radiographic testing

Personnel performing radiography examination to this procedure shall be qualified and certified by XXX also shall meet the requirements of ASNT-SNT-TC-1A-2001 EDITION at least level Ⅱ and on ASNT-SNT-TC-IA for code section Ⅰ and Sec. Ⅶ, Div. Ⅱ. Film interpreter shall have level Ⅱ as a minimum.

Item 6

06 Ultrasonic testing

Learning Objectives

1. Knowledge objectives
(1) To grasp the words, related terms and abbreviations about UT.

(2) To grasp the classification about UT system.

(3) To know the instruments and equipment of UT system.

(4) To know the testing procedure of UT system.

2. Competence objectives
(1) To be able to read and understand frequently used & complex sentence patterns, capitalized English materials and obtain key information quickly.

(2) To be able to communicate with English speakers about the topic freely.

(3) To be able to fill in the job cards in English.

3. Quality objectives
(1) To be able to self-study with the help of aviation dictionaries, the Internet and other resources.

(2) To do a good job of detection of safety protection.

6.1 Introduction of Ultrasonic Testing

Ultrasonic testing (Fig. 6–1) is the use of high frequency sound to inspect materials. The sound frequency normally ranges from 5 to 25 MHz, which is well above what a human can hear. High frequency sound is used because it is able to penetrate most materials without difficulty, which makes it possible to conduct non-destructive testing. It is commonly used to look for cracks, measure material depth, and check for corrosion and imperfections.

An ultrasonic test works by

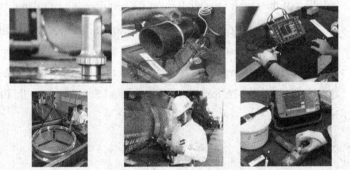

Fig. 6–1 Ultrasonic Testing

bouncing sound off of an object and interpreting the resulting echo. It passes through the material until the sound waves encounter an irregularity. The operators usually note this as a "discontinuity". By analyzing the discontinuity, the operator can determine if there is a flaw in the material.

The thickness of materials such as metals, ceramics and plastics can be measured with ultrasonic testing. Ultrasonic thickness testing is mainly done by calculating the time it takes for sound to bounce off the bottom of the material. Different materials typically reflect sound at different rates. By measuring the change in the time it takes for the sound to be reflected, the operator is able to measure the thickness of each material in a multilayer surface.

There are several types of ultrasonic testing equipment, depending on the required application. The choice of equipment is generally dictated by the material's temperature, thickness, geometry and phase reversal. Ultrasonic test equipment typically has three components: a transducer, a couplant and an imaging system. A technician operates the ultrasonic testing equipment by manually moving the probe across the surface of the object being tested and interpreting the resulting data.

A transducer or probe produces and receives sound. Normally, a transducer sends sound in either a straight beam or in an angle beam. Straight beam transducers are more widely used than angle beam transducers, which are often used for ultrasonic weld testing.

Data from the test can be read with an imaging system. The imaging system normally contains the controls and processor. In some portable equipment, the transducer is also integrated into the imaging system.

■ 〖Point 1〗 Basic Principles of Ultrasonic Testing

Ultrasonic Testing (UT) uses high frequency sound energy to conduct examinations and make measurements. Ultrasonic inspection can be used for flaw detection/evaluation, dimensional measurements, material characterization, and more. To illustrate the general inspection principle, a typical pulse/echo inspection configuration as illustrated below will be used (Fig. 6–2).

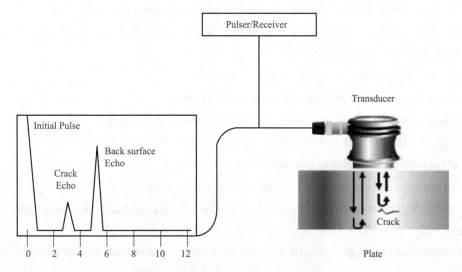

Fig. 6–2 Typical UT Inspection System

A typical UT inspection system consists of several functional units, such as the pulser/receiver, transducer, and display devices. A pulser/receiver is an electronic device that can produce high voltage electrical pulses. Driven by the pulser, the transducer generates high frequency ultrasonic energy. The sound energy is introduced and propagates through the materials in the form of waves. When there is a discontinuity (such as a crack) in the wave path, part of the energy will be reflected back from the flaw surface. The reflected wave signal is transformed into an electrical signal by the transducer and is displayed on a screen (Fig. 6–3). In the applet below, the reflected signal strength is displayed versus the time from signal generation to when a echo was received. Signal travel time can be directly related to the distance that the signal traveled. From the signal, information about the reflector location, size, orientation and other features can sometimes be gained.

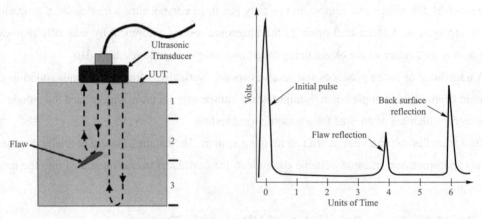

Fig. 6–3 The Transducer and Screen

Ultrasonic Inspection is a very useful and versatile NDT method. Some of the advantages of ultrasonic inspection that are often cited include:
- It is sensitive to both surface and subsurface discontinuities.
- The depth of penetration for flaw detection or measurement is superior to other NDT methods.
- Only single-sided access is needed when the pulse-echo technique is used.
- It is highly accurate in determining reflector position and estimating size and shape.
- Minimal part preparation is required.
- Electronic equipment provides instantaneous results.
- Detailed images can be produced with automated systems.
- It has other uses, such as thickness measurement, in addition to flaw detection.

As with all NDT methods, ultrasonic inspection also has its limitations, which include:
- Surface must be accessible to transmit ultrasound.
- Skill and training is more extensive than with some other methods.
- It normally requires a coupling medium to promote the transfer of sound energy into the test specimen.
- Materials that are rough, irregular in shape, very small, exceptionally thin or not homogeneous are difficult to inspect.

• Cast iron and other coarse grained materials are difficult to inspect due to low sound transmission and high signal noise.

• Linear defects oriented parallel to the sound beam may go undetected.

• Reference standards are required for both equipment calibration and the characterization of flaws.

The above introduction provides a simplified introduction to the NDT method of ultrasonic testing. However, to effectively perform an inspection using ultrasonics, much more about the method needs to be known. The following pages present information on the science involved in ultrasonic inspection, the equipment that is commonly used, some of the measurement techniques used, as well as other information.

■ 〖Point 2〗 History of Ultrasonics

The birth of the modern science and technology of utrasonics can be placed in 1917, when Langevin first achieved real success in his attempts to detect the presence of submarines by echo ranging with ultrasound (Hunt, 1954) . The idea had been suggested some time earlier: the Titanic disaster in 1912 prompted Richardson to file a patent describing the method, but the techniques available then were inadequate. Langevin began his work in 1915 with electrostatic transducers to launch and receive the ultrasound; in 1917 he first used an X-cut quartz plate, about 10 cm × 10 cm × 1.5 cm, cut from a large and beautiful.

Quartz crystal which had graced the window of a Paris optician. Such large crystals are rare; in the following year Langevin successfully used a mosaic of small pieces of quartz sandwiched between steel plates and also experimented with magnetostrictors. Langevin's work led directly to the modern asdic and sonar systems and the closely related applications to depth sounding and fish-finding. The same principle, timing the go and return flight of a pulse of ultrasound, is applied in solid materials to detect flaws. Thickness gauging uses a continuous-wave embodiment of the same princple of the reflection of ultrasonic waves. The most recently developed applications of the pulse echo principle are to biological subjects. The thickness of the back-fat of live pigs is now being measured in this way. In the human subject (Gordon, 1963), various individual organs—the brain, the eye, the breast—can be examined for the presence of abnormal structure or foreigr, bodies. Complex variations on the same theme, involving multiple scanning from many points on the skin surface and in many directions from each point, coupled with integration of all the echo information by photography, permit an "echo cross section" of the neck, a limb or even the abdomen to be built up. These ultrasonic aids to medical diagnosis supplement the established methods such as X-ray photography. They are still undergoing development in both the U.K and U.S.A. They have already shown such promise that it has been proposed in U.S.A. to construct automatic ultrasonic echo scanning apparatus for "mass" use, like mass radio-graphy, in searching for cancer of the breast. The ultrasonic method is reported to detect this disease at a much earlier stage than any other method, so offering the best chance of successful surgical treatment. All these pulse echo systems are based on the difference in reflecting properties of different parts of the subject under examination. Another approach to the visualization of internal structure is based on the difference in the trans-mission properties. Ultrasound is refracted at the interface between two media in which its velocity has different values. It can therefore be focused by "lenses" and "mirrors" and image forming systems can be constructed. In the ultrasonoscope, an image of the object to he examined is formed on a plate of piezoelectric material. At each point, the intensity of the sonic image produces a corresponding electrical polarization of the plate, which transforms the image into a pattern of electrical charge. The plate is mounted on the end of an evacuated tube which also contains an electron gun arranged to scan the plate with a beam of electrons. This generates an electrical signal which can be applied to a normal cathode-ray tube, synchronously scanned, to create a visible image of the original object in terms of its opacity to the ultrasonic wave. This system is still in an early experimental stage and will require considerable technological development before its ultimate value can be assessed. A landmark in the history of high-power applications was found in 1927, when Wood and Loomis (1927) published the description of the effects they had produced with intense ultrasound at 300 kc/s. They applied 50 kV from a 2 kW valve oscillator to a quartz crystal immersed in oil. The radiation pressure on a disk 8 cm in diameter would support a mass of 150 g. A mound 7 cm high was raised in the oil with a fountain from which drops were thrown to a distance of 40 cm. By concentrating the ultrasound in tapering glass rods and drawn-down tubes, they showed that the vibration could drill wood (by burning) and glass (by chipping and fusion), emulsify oil and water, atomize oil or mercury and cause either flocculation or dispersion in suspensions of solids in liquids. All these effects have been developed and applied in industry during the last 20 years. Only one major effect has been added to those demonstrated by Wood and Loomis: during the Second World War German workers seeking to improve the quality of electrical resistance welds discovered that ultrasonic vibration alone would weld certain metals together.

■ 〖Point 3〗 Development of Ultrasonics

Early ultrasonic flaw detectors (Fig. 6-4) used vacuum tubes (valves), needed generated electricity, and were heavy. Using quartzcrystals, signal amplitude was poor and resolution very poor. After a shaky start, semiconductor technology has produced flaw detectors that are light, very portable, and together with synthetic crystal materials offers performance that is greatly enhanced.

Early ultrasonic instrument called the Supersonic Reflectoscope

Fig. 6-4 Early Ultrasonic Flaw Detectors

Much of this had been achieved by the mid 1970s. During the 1980s and 1990s, microchips have been incorporated into the flaw detector, allowing the operator to store calibration parameters and signal traces. This, in turn, allows off-line analysis and reevaluation at a later date. Digital technology and the use of LCD display panels instead of CRTs during the 1990s has further reduced the size and weight of the flaw detectors.

Phased Array Ultrasonic Testing (Fig. 6-5) is an advanced application of ultrasonic testing technology. It can be used for weld inspections, crack and flaw detections, thickness measurements and corrosion inspections. Because of the detailed visualisation of the defect size, shape, depth and orientation, Phased Array can often be used instead of Radiographic Testing. Since it doesn't use ionising radiation, there is no need to create a safety zone which usually means interrupting production.

In conventional UT testing, a single transducer sends ultrasound waves into the material. Phased Array probes contain multiple transducers. By introducing a delay between the pulses sent out by each transducer, the beam angle, focal point and focal spot of the generated wavefront can be influenced. This makes Phased Array Ultrasonic Testing a very versatile method that can be used for complex geometries.

The importance of welding in structures across all sectors of industry can't be overstated. Determining the quality of welds is often vital to the longevity and profitability of structures. Time-of-Flight Diffraction (TOFD) (Fig. 6–5) is an advanced NDT inspection method which is widely used for weld testing. Although not exclusively used for welds, it is very effective in detecting defects such as cracks, lack of fusion and slag inclusion.

TOFD is an innovation of ultrasonic testing which uses two probes. One probe sends the ultrasound and the other one is used as a receiver. Instead of recording the ultrasound that is reflected by defects, it detects diffraction of sound waves that emanate from the tips of a defect. TOFD inspection provides accurate information about the length and height of defects. Unlike other testing methods, TOFD can detect flaws regardless of their orientation.

Phased Array
Ultrasonic Testing (PA)

Time-of-Flight
Diffraction (TOFD)

Fig. 6–5 Advanced Ultrasonic Flaw Detectors

■ 〖Point 4〗 Benefits and Applications

Ultrasonic inspection is most often performed on steels and metals. It can be used on other nonmetallic materials with success. This form of non-destructive testing is widely used many industries including aerospace, automotive, pressure vessel, and the welding fabrication industry. It can locate subsurface discontinuities in weldments when access to only one side is possible. It has many advantages such as: high penetrating power for going thru very thick parts, high sensitivity for the detection of small discontinuities, only one surface need be accessible, non hazardous to the operator or to nearby personnel, and it is highly portable.

Benefits of Ultrasonic NDT
- Usually requires one surface open for inspection
- Cost effective
- Portable
- Inspection of both ferrous and non-ferrous materials
- Fast results
- Parts can be inspected in a production atmosphere

Ultrasonic testing is used in aerospace, automotive and transportation industries, it often performed on steel and other alloys. For certain industries, such as aerospace, the materials used may be thinner than most with tolereance thresholds that require these high precision measurements. It plays an important role in safety, quality assurance and cost.

Applications of UT Testing
- Forgings
- Weldments
- Castings
- Bars and shapes
- Thickness gauging
- Velocity/Nodularity
- Plate inspection
- Pipes/Tubes

Put into Practice

1. Translate the content in Fig. 6-6.

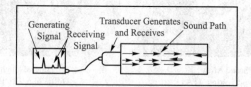

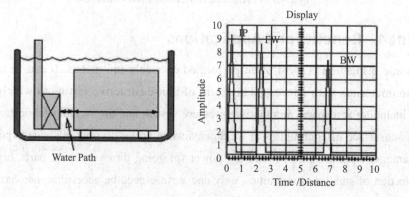

Fig. 6-6 Display

2. The following ultrasound detection system diagrams are translated (Fig. 6-7).

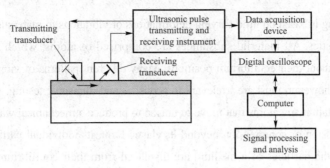

Fig. 6-7 Ultrasound Detection System

6.2　Physics of Ultrasound

〖Point 1〗The Nature of Ultrasound

Sound waves are mechanical vibrations that propagate in a host medium. They are coupled modes between medium particles oscillating about equilibrium positions and a traveling ultrasonic wave. Solids support the propagation of both longitudinal waves (particles oscillating parallel to the wave propagation direction) and transverse waves (particles oscillating perpendicular to the wave propagation direction). Fluids (gases and liquids) only support longitudinal wave propagation. The lean body mass is approximately 72% water and the remainder is fat, which is fluidlike, so only longitudinal waves can be used to probe the human body. Transverse waves may be generated in bone due to mode conversion, but because of the bone's high attenuation, they do not contribute to ultrasonic image formation.

Propagating sound waves obey the standard relation:

$$c=\lambda f$$

Where c is the acoustic velocity in the medium, f is the frequency of the wave, and λ is the acoustic wavelength. For single-frequency, continuous wave (CW) sound waves, at a single point in the medium f is the number of incident pressure (or any other wave parameter) cycles per second (Hz) and at a single instant of time λ is the basic spatial cyclic repetition distance of the single-frequency wave.

Sound waves with frequencies above 1 MHz (ultrasound) can be easily generated and focused and will propagate reasonable distances in soft tissue. Pulse-echo ultrasonic measurements, similar to those of sonar and radar, are now used routinely in medicine to provide detailed images of cross-sectional anatomy.

This section of the chapter will review the basic physical principles used in medical ultrasonic image formation. Only essential principles will be described; details are available in the references.

〔Point 2〕 Wave Propagation

Ultrasonic testing is based on time-varying deformations or vibrations in materials, which is generally referred to as acoustics. All material substances are comprised of atoms, which may be forced into vibrational motion about their equilibrium positions. Many different patterns of vibrational motion exist at the atomic level, however, most are irrelevant to acoustics and ultrasonic testing. Acoustics is focused on particles that contain many atoms that move in unison to produce a mechanical wave. When a material is not stressed in tension or compression beyond its elastic limit, its individual particles perform elastic oscillations. When the particles of a medium are displaced from their equilibrium positions, internal (electrostatic) restoration forces arise. It is these elastic restoring forces between particles, combined with inertia of the particles, that leads to the oscillatory motions of the medium.

Sound propagates through air as a longitudinal wave. The speed of sound is determined by the properties of the air, and not by the frequency or amplitude of the sound. Sound waves (Fig. 6-8), as well as most other types of waves, can be described in terms of the following basic wave phenomena.

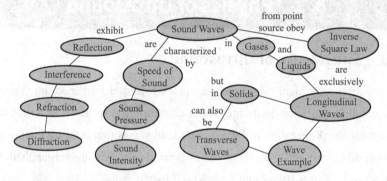

Fig. 6-8 Sound Wave

Longitudinal waves (Fig. 6-9) can propagate through either solid or fluid media, but transverse or "shear" waves cannot propagate in a fluid. This can be used to advantage in geological studies as depicted in the illustration below.

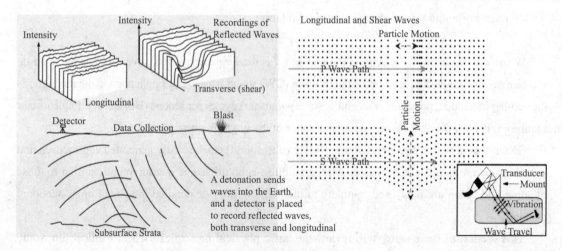

Fig. 6-9 Longitudinal Wave

Since the longitudinal waves travel through both solid and liquid, the longitudinal data can be used as a reference for mapping the transverse wave data. Collecting seismic data from both longitudinal waves and transverse waves and taking the difference allows the experimenters to map the underground liquified region.

Both types of waves in the material of the earth are known as seismic waves. The longitudinal waves are called P waves and the transverse waves S waves. P waves typically travel almost twice as fast as S waves.

When ground movement like that recorded in the Mammoth Lakes region of California is discovered, then this kind of study can give some idea about the intrusion of magma underneath the surface.

■ 〖Point 3〗 Modes of Sound Wave Propagation

In air, sound travels by the compression and rarefaction of air molecules in the direction of travel. However, in solids, molecules can support vibrations in other directions, hence, a number of different types of sound waves are possible. Waves can be characterized in space by oscillatory patterns that are capable of maintaining their shape and propagating in a stable manner. The propagation of waves is often described in terms of what are called "wave modes".

As mentioned previously, longitudinal and transverse (shear) waves are most often used in ultrasonic inspection. However, at surfaces and interfaces, various types of elliptical or complex vibrations of the particles make other waves possible. Some of these wave modes such as Rayleigh and Lamb waves are also useful for ultrasonic inspection.

Table 6-1 summarizes many, but not all, of the wave modes possible in solids.

Table 6-1 Wave Types in Solids

Wave Types in Solids	Particle Vibrations
Longitudinal	Parallel to wave direction
Transverse (Shear)	Perpendicular to wave direction
Surface-Rayleigh	Elliptical orbit-symmetrical mode
Plate Wave-Lamb	Component perpendicutar to surface (extensional wave)
Plate Wave-Love	Parallel to plane layer, perpendicular to wave direction
Stoneley (Leaky Rayleigh Waves)	Wave guided along interface
Sezawa	Antisymmetric mode

Longitudinal and transverse waves were discussed on the previous page, so let's touch on surface and plate waves here.

Surface (or Rayleigh) waves (Fig. 6-10) travel the surface of a relatively thick solid material penetrating to a depth of one wavelength. Surface waves combine both a longitudinal and transverse motion to create an elliptic orbit motion as shown in the image and animation below. The major axis of the ellipse is perpendicular to the surface of the solid. As the depth of an individual atom from the surface increases the width of its elliptical motion decreases. Surface waves are generated when a longitudinal wave intersects a surface near the second critical angle and they travel at a velocity between 0.87 and 0.95 of a shear wave. Rayleigh waves are useful because they are very sensitive to surface defects (and other surface features) and they follow the surface around curves. Because of this, Rayleigh waves can be used to inspect areas that other waves might have difficulty reaching.

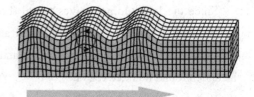

Fig. 6–10 Surface (or Rayleigh) Waves

■ 〖Point 4〗 Wavelength, Frequency and Velocity

Among the properties of waves propagating in isotropic solid materials are wavelength, frequency, and velocity. The wavelength is directly proportional to the velocity of the wave and inversely proportional to the frequency of the wave. This relationship is shown by the following equation.

$$\text{Wavelength}(\lambda) = \frac{\text{Velocity}(v)}{\text{Frequency}(f)}$$

The applet below shows a longitudinal and transverse wave. The direction of wave propagation is from left to right and the movement of the lines indicate the direction of particle oscillation. The equation relating ultrasonic wavelength, frequency, and propagation velocity is included at the bottom of the applet in a reorganized form. The values for the wavelength, frequency, and wave velocity can be adjusted in the dialog boxes to see their effects on the wave. Note that the frequency value must be kept between 0.1 to 1 MHz (one million cycles per second) and the wave velocity must be between 0.1 and 0.7 cm/μs.

As can be noted by the equation, a change in frequency will result in a change in wavelength. Change the frequency in the applet and view the resultant wavelength. At a frequency of 0.2 and a material velocity of 0.585 (longitudinal wave in steel) note the resulting wavelength. Adjust the material velocity to 0.480 (longitudinal wave in cast iron) and note the resulting wavelength. Increase the frequency to 0.8 and note the shortened wavelength in each material.

In ultrasonic testing, the shorter wavelength resulting from an increase in frequency will usually provide for the detection of smaller discontinuities. This will be discussed more in following sections.

■ 〖Point 5〗 The Speed of Sound

Of course, sound does travel at different speeds in different materials. This is because the mass of the atomic particles and the spring constants are different for different materials. The mass of the particles is related to the density of the material, and the spring constant is related to the elastic constants of a material. The general relationship between the speed of sound in a solid and its density and elastic constants is given by the following equation:

$$V = \sqrt{\frac{C_{ij}}{\rho}}$$

Where V is the speed of sound, C is the elastic constant, and ρ is the material density. This equation may take a number of different forms depending on the type of wave (longitudinal or shear) and which of the elastic constants that are used. The typical elastic constants of a materials include:

• Young's Modulus, E: a proportionality constant between uniaxial stress and strain.
• Poisson's Ratio, n: the ratio of radial strain to axial strain.
• Bulk modulus, K: a measure of the incompressibility of a body subjected to hydrostatic pressure.
• Shear Modulus, G: also called rigidity, a measure of a substance's resistance to shear.
• Lame's Constants, l and m: material constants that are derived from Young's Modulus and Poisson's Ratio.

When calculating the velocity of a longitudinal wave, Young's Modulus and Poisson's Ratio are commonly used. When calculating the velocity of a shear wave, the shear modulus is used. It is often most convenient to make the calculations using Lame's Constants, which are derived from Young's Modulus and Poisson's Ratio.

It must also be mentioned that the subscript ij attached to C in the above equation is used to indicate the directionality of the elastic constants with respect to the wave type and direction of wave travel. In isotropic materials, the elastic constants are the same for all directions within the material. However, most materials are anisotropic and the elastic constants differ with each direction. For example, in a piece of rolled aluminum plate, the grains are elongated in one direction and compressed in the others and the elastic constants for the longitudinal direction are different than those for the transverse or short transverse directions.

■ 〖Point 6〗 Attenuation of Sound Waves

When a wave travels through a medium, its intensity diminishes with distance. In idealized materials, the wave amplitude is only reduced by the spreading of the wave. Natural materials, however, all produce an effect which further weakens the wave. This further weakening results from scattering and absorption. Scattering is the reflection of the wave in directions other than its original direction of propagation. Absorption is the conversion of the wave energy to other forms of energy. The combined effect of scattering and absorption is called attenuation (Fig. 6–11).

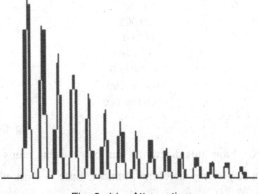

Fig. 6–11 Attenuation

The amplitude change of a decaying/attenuating plane wave can be expressed as:

$$A = A_0 e^{-\alpha z}$$

In this expression A_0 is the unattenuated amplitude of the propagating wave at some location. The amplitude A is the reduced amplitude after the wave has traveled a distance z from that initial location. The quantity α/alpha is the attenuation coefficient of the wave traveling in the z-direction. The dimensions of α/alpha are nepers/length, where a neper is a dimensionless quantity. The term e is the exponential (or Napier's constant) which is equal to approximately 2.718 28.

The units of the attenuation value in Nepers per meter (Np/m) can be converted to decibels/length by dividing by 0.115 1. Decibels is a more common unit when relating the amplitudes of two signals. More information about decibels is provided below.

The Decibel

The decibel (dB) is one tenth of a Bel, which is a unit of measure that was developed by engineers at Bell Telephone Laboratories and named for Alexander Graham Bell. The dB is a logarithmic unit that describes a ratio of two measurements. The basic equation that describes the difference in decibels between two measurements is:

$$\Delta X (\mathrm{dB}) = 10 \log \frac{X_2}{X_1}$$

Where ΔX is the difference in some quantity expressed in decibels, X_1 and X_2 are two different measured values of X, and the log is to base 10.

Why is the dB unit used?

Use of dB units allows ratios of various sizes to be described using easy to work with numbers. For example, consider the information in Fig. 6–12.

Ratio between Measurement 1 and 2	Equation	dB
1/2	dB=10log (1/2)	−3 dB
1	dB=10log (1)	0 dB
2	dB=10log (2)	3 dB
10	dB=10log (10)	10 dB
100	dB=10log (100)	20 dB
1 000	dB=10log (1 000)	30 dB
10 000	dB=10log (10 000)	40 dB
100 000	dB=10log (100 000)	50 dB
1 000 000	dB=10log (1 000 000)	60 dB
10 000 000	dB=10log (10 000 000)	70 dB
100 000 000	dB=10log (100 000 000)	80 dB
1 000 000 000	dB=10log (1 000 000 000)	90 dB

Fig. 6–12 dB

From this diagram it can be seen that ratios from one up to ten billion can be represented with a single or double digit number. Ease to work with numbers was particularly important in the days before the advent of the calculator or computer. The focus of this discussion is on using the dB in measuring sound levels, but it is also widely used when measuring power, pressure, voltage and a number of other things.

■ 〖Point 7〗 Wave Interaction or Interference

Wave propagation has been discussed so far as if a single sinusoidal wave was propagating through the material. However, the sound that emanates from an ultrasonic transducer or the EM waves that radiate from an antenna do not originate from a single point, but instead originates from many points. This results in a field with many waves interacting or interfering with each other.

When waves interact, they superimpose on each other, and the amplitude of the sound pressure or particle displacement at any point of interaction is the sum of the amplitudes of the two individual waves. First, let's consider two identical waves that originate from the same point. When they are in phase (so that the peaks and valleys of one are exactly aligned with those of the other), they combine to double the displacement of either wave acting alone. This is the case on the left and this is called constructive interference. When they are completely out of phase (so that the peaks of one wave are exactly aligned with the valleys of the other wave), they combine to cancel each other out. This is the case in the middle and this is a case of destructive interference. When the two waves are not completely in phase or out of phase, the resulting wave is the sum of the wave amplitudes for all points along the wave. This is the case on the right and in this case there is both constructive and destructive interference.

When the origins of the two interacting waves are not the same, it is a little harder to picture the wave interaction, but the principles are the same. Up until now, we have primarily looked at waves in the form of a 2D plot of wave amplitude versus wave position. However, anyone that has dropped something in a pool of water can picture the waves radiating out from the source with a circular wave front. If two objects are dropped a short distance apart into the pool of water, their waves will radiate out from their sources and interact with each other. At every point where the waves interact, the amplitude of the particle displacement is the combined sum of the amplitudes of the particle displacement of the individual waves.

With an ultrasonic transducer or an antenna, the waves propagate out from the transducer face with a circular wave front. If it were possible to get the waves to propagate out from a single point on the transducer face, the field would appear as shown in the upper image to the right. Consider the light areas to be areas of rarefaction and the dark areas to be areas of compression.

However, as stated previously, waves originate from multiple points along the face of the transducer or antenna (Fig. 6–13). The lower image to the right shows what the field would look like if the waves originated from just two points. It can be seen that where the waves interact, there are areas of constructive and destructive interference. The points of constructive interference are often referred to as nodes. Of course, there are more than two points of origin along the face of a transducer or antenna. The image below shows five points of wave origination. It can be seen that near the face of the transducer, there are extensive fluctuations or nodes and the field is very uneven. In ultrasonic testing and in antenna theory, this in known as the near field (near zone) or Fresnel zone. The field is more uniform away from the transducer in the far field, or Fraunhofer zone, where the beam spreads out in a pattern originating from the center of the transducer/antenna. It should be noted that even in the far field, it is not a uniform wave front. However, at some distance from the face of the transducer and central to the face of the transducer, a uniform and intense wave field develops.

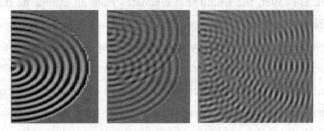

Fig. 6-13 Waves Interact

In the image to the left, multiple points of sound origination along the face of the transducer/antenna.

The curvature and the area over which the wave is being generated, the speed that the waves travel within a material and the frequency of the wave all affect the sound field. Use the Java applet below to experiment with these variables and see how the sound field is affected.

The black and blue lines represent longitudinal wave axis. The red line represents the mode converted shear wave. Users may change materials, velocities and angles.

■〖Point 8〗 Refraction and Snell's Law

When an ultrasonic or EM wave passes through an interface between two materials at an oblique angle, and the materials have different indices of refraction, both reflected and refracted waves are produced. This also occurs with light, which is why objects seen across an interface appear to be shifted relative to where they really are. For example, if you look straight down at an object at the bottom of a glass of water, it looks closer than it really is. A good way to visualize how light and sound refract is to shine a flashlight into a bowl of slightly cloudy water noting the refraction angle with respect to the incident angle.

Oblique Angle: an angle that is not a right angle or a multiple of a right angle.

Refraction: a change in direction of a wave due to a change in its speed.

Refraction takes place at an interface due to the different velocities of the waves within the two materials. The velocity of the wave in each material is determined by the material properties (elastic modulus, density, dielectric properties, refractive index) for that material. In the animation below, a series of plane waves are shown traveling in one material and entering a second material that has a higher wave speed. Therefore, when the wave encounters the interface between these two materials, the portion of the wave in the second material is moving faster than the portion of the wave in the first material. It can be seen that this causes the wave to bend.

Snell's Law (Fig. 6-14) describes the relationship between the angles and the velocities of the waves. Snell's law equates the ratio of material velocities V_1 and V_2 to the ratio of the sine's of incident (θ_1) and refracted (θ_2) angles, as shown in the following equation.

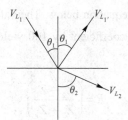

Fig. 6-14 Snell's Law

$$\frac{\sin\theta_1}{V_{L_1}} = \frac{\sin\theta_2}{V_{L_2}}$$

Where: V_{L_1}—the longitudinal wave velocity in material 1;
V_{L_2}—the longitudinal wave velocity in material 2.

Note that in the diagram above, there is a reflected longitudinal wave (V_{L_1}) shown. This wave is reflected at the same angle as the incident wave because the two waves are traveling in the same material, and hence have the same velocities. This reflected wave is unimportant in our explanation of Snell's Law, but it should be remembered that some of the wave energy is reflected at the interface. In the applet below, only the incident and refracted waves are shown. The angle of either wave can be adjusted by clicking and dragging the mouse in the region of the arrows. Values for the angles or wave velocities can also be entered in the dialog boxes so the that applet can be used as a Snell's Law calculator.

When a wave moves from a slower to a faster material, there is an incident angle that makes the angle of refraction for the wave 90°. This is knowing as the first critical angle. The first critical angle can be found from Snell's law by putting in an angle of 90° for the angle of the refracted ray. In the case of acoustic waves, at the critical angle of incidence much of the acoustic energy is in the form of an inhomogeneous compression wave, which travels along the interface and decays exponentially with depth from the interface. This wave is sometimes referred to as a "creep wave". Because of their inhomogeneous nature and the fact that they decay rapidly, creep waves are not used as extensively as rayleigh surface waves in NDT. However, creep waves are sometimes more useful than rayleigh waves because they suffer less from surface irregularities and coarse material microstructure due to their longer wavelengths.

【Point 9】 Reflection and Transmission Coefficients

Waves are reflected at boundaries where there is a difference in impedances (Z) of the materials on each side of the boundary. This difference in Z is commonly referred to as the impedance mismatch. The greater the impedance mismatch, the greater the percentage of energy that will be reflected at the interface or boundary between one medium and another.

The fraction of the incident wave intensity that is reflected can be derived because particle velocity and local particle pressures must be continuous across the boundary. When the impedances of the materials on both sides of the boundary are known, the fraction of the incident wave intensity that is reflected can be calculated with the equation below. The value produced is known as the reflection coefficient. Multiplying the reflection coefficient by 100 yields the amount of energy reflected as a percentage of the original energy.

$$R = \left(\frac{Z_2 - Z_1}{Z_2 + Z_1}\right)^2.$$

Since the amount of reflected energy plus the transmitted energy must equal the total amount of incident energy, the transmission coefficient is calculated by simply subtracting the reflection coefficient from one.

Coefficients: A calculated measure of a physical property such as reflected or transmitted waves.

Formulations for reflection and transmission coefficients (pressure) are shown in the interactive applet below. Different materials may be selected or the material velocity and density may be altered to change the acoustic impedance of one or both materials. The red arrow represents reflected sound and the blue arrow represents transmitted sound.

■ 〖Point 10〗 Mode Conversion (Fig. 6–15)

When a wave travels in a solid material, one form of wave energy can be transformed into another form. For example, when a longitudinal waves hits an interface at an angle, some of the energy can cause particle movement in the transverse direction to start a shear (transverse) wave. Mode conversion occurs when a wave encounters an interface between materials of different impedances and the incident angle is not normal to the interface. From the ray tracing clip below, it can be seen that since mode conversion occurs every time a wave encounters an interface at an angle, signals can become confusing at times.

Material properties: Material strain is proportional to material stress.

Elastic modulus: The ratio of the stress applied to a body to the strain produced.

In the previous sections, it was pointed out that when waves pass through an interface between materials having different wave speeds, refraction takes place at the interface. The larger the difference in wave speeds between the two materials, the more the wave is refracted. Notice that the shear wave is not refracted as much as the longitudinal wave. This occurs because shear waves travel slower than longitudinal waves. Therefore, the velocity difference between the incident longitudinal wave and the shear wave is not as great as it is between the incident and refracted longitudinal waves. Also note that when a longitudinal wave is reflected inside the material, the reflected shear wave

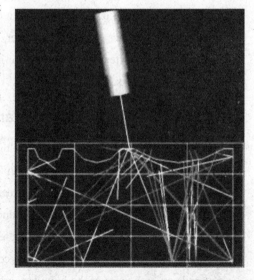

Fig. 6–15 Mode Conversion

is reflected at a smaller angle than the reflected longitudinal wave. This is also due to the fact that the shear velocity is less than the longitudinal velocity within a given material.

Snell's Law (Fig. 6-16) holds true for shear waves as well as longitudinal waves and can be written as follows.

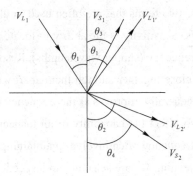

Fig. 6-16 Snell's Law

$$\frac{\sin\theta_1}{V_{L_1}} = \frac{\sin\theta_2}{V_{L_2}} = \frac{\sin\theta_3}{V_{S_1}} = \frac{\sin\theta_4}{V_{S_2}}$$

Where: V_{L_1}—the longitudinal wave velocity in material 1;
V_{L_2}—the longitudinal wave velocity in material 2;
V_{S_1}—the shear wave velocity in material 1;
V_{S_2}—the shear wave velocity in material 2.

In the applet below, the shear (transverse) wave ray path has been added. The ray paths of the waves can be adjusted by clicking and dragging in the vicinity of the arrows. Values for the angles or the wave velocities can also be entered into the dialog boxes. It can be seen from the applet that when a wave moves from a slower to a faster material, there is an incident angle which makes the angle of refraction for the longitudinal wave 90 degrees. As mentioned on the previous page, this is known as the first critical angle and all of the energy from the refracted longitudinal wave is now converted to a surface following longitudinal wave. This surface following wave is sometime referred to as a creep wave and it is not very useful in NDT because it dampens out very rapidly.

Beyond the first critical angle, only the shear wave propagates into the material. For this reason, most angle beam transducers use a shear wave so that the signal is not complicated by having two waves present. In many cases there is also an incident angle that makes the angle of refraction for the shear wave 90 degrees. This is known as the second critical angle and at this point, all of the wave energy is reflected or refracted into a surface following shear wave or shear creep wave. Slightly beyond the second critical angle, surface waves will be generated.

■ 〖Point 11〗 Wavelength and Defect Detection

In ultrasonic and microwave NDT, the inspector must make a decision about the frequency of the transducer that will be used. As we learned on previously, changing the frequency when the wave velocity is fixed will result in a change in the wavelength of the sound. The wavelength used has a significant effect on the probability of detecting a discontinuity. A general rule of thumb is that a discontinuity must be larger than one-half the wavelength to stand a reasonable chance of being detected.

Sensitivity and resolution are two terms that are often used in ultrasonic inspection to describe a technique's ability to locate flaws. Sensitivity is the ability to locate small discontinuities. Sensitivity generally increases with higher frequency (shorter wavelengths). Resolution is the ability of the system to locate discontinuities that are close together and tell them apart within the material or located near the part surface. Resolution also generally improves as the frequency increases.

The wave frequency can also affect the capability of an inspection in adverse ways. Therefore, selecting the optimal inspection frequency often involves maintaining a balance between the favorable and unfavorable results of the selection. Before selecting an inspection frequency, the material's grain structure and thickness, and the discontinuity's type, size, and probable location should be considered. As frequency increases, waves tend to scatter from large or course grain structure and from small imperfections within a material. Cast materials often have coarse grains and other sound scatters that require lower frequencies to be used for evaluations of these products. Wrought and forged products with directional and refined grain structure can usually be inspected with higher frequency transducers.

Since more things in a material are likely to scatter a portion of the wave energy at higher frequencies, the penetrating power (or the maximum depth in a material that flaws can be located) is also reduced.

■〖Point 12〗 Signal-to-Noise Ratio

In a previous page, the effect that frequency and wavelength have on flaw detectability was discussed. However, the detection of a defect involves many factors other than the relationship of wavelength and flaw size. For example, the amount of the wave that reflects from a defect is also dependent on the impedance mismatch between the flaw and the surrounding material. A void is generally a better reflector than a metallic inclusion because the impedance mismatch is greater between air and metal than between two metals.

Often, the surrounding material has competing reflections. Microstructure grains in metals and the aggregate of concrete are a couple of examples. A good measure of detectability of a flaw is its signal-to-noise ratio (S/N). The signal-to-noise ratio is a measure of how the signal from the defect compares to other background reflections (categorized as "noise"). A signal-to-noise ratio of 3 to 1 is often required as a minimum. The absolute noise level and the absolute strength of an echo from a "small" defect depends on a number of factors, which include:

The probe size and focal properties.

The probe frequency, bandwidth and efficiency.

The inspection path and distance (water/solid).

The interface (surface curvature and roughness).

The flaw location with respect to the incident beam.

The inherent noisiness of the metal microstructure.

The inherent reflectivity of the flaw, which is dependent on its acoustic impedance, size, shape and orientation.

Cracks and volumetric defects can reflect ultrasonic waves quite differently. Many cracks are "invisible" from one direction and strong reflectors from another.

Multifaceted flaws will tend to scatter sound away from the transducer.

Put into Practice

1. Fig. 6-17 analyzes the absorption and scattering of different materials.

Nature of Material	Attenuation* (dB/m)	Principal
Normalized Steel	70	Scatter
Aluminum, 6061-T6511	90	Scatter
Stainless Steel, 3XX	110	Scatter/Redirection
Plastic (clear acrylic)	380	Absorption
*Frequency of 2.25 MHz, Longitudinal wave mode		

Fig. 6-17 The Absorption and Scattering

2. Translate the content in Fig. 6-18.

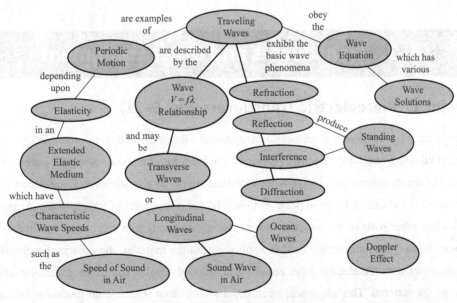

Fig. 6-18 The Chain Diagram

3. Look at Fig. 6-19 and point out the frequency range of the sound wave and the frequency range of the ultrasonic wave used in the test.

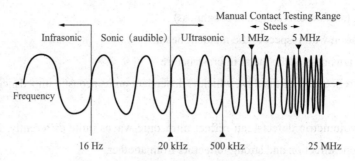

Fig. 6-19 Frequency Range

4. Fig. 6-20 is translated and the propagation characteristics of ultrasonic waves are explained.

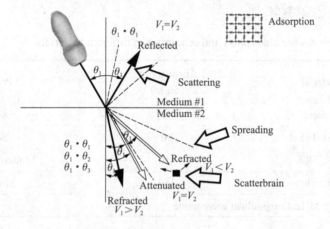

Fig. 6-20 The Propagation Characteristics

6.3 Transducers and Other Equipment

〖Point 1〗 Piezoelectric Transducers (Fig. 6-21)

The conversion of electrical pulses to mechanical vibrations and the conversion of returned mechanical vibrations back into electrical energy is the basis for ultrasonic testing. The active element is the heart of the transducer as it converts the electrical energy to acoustic energy, and vice versa. The active element is basically a piece of polarized material (i.e. some parts of the molecule are positively charged, while other parts of the molecule are negatively charged) with electrodes attached to two of its opposite faces. When an electric field is applied across the material, the polarized molecules will align themselves with the electric field, resulting in induced dipoles within the molecular or crystal structure of the material. This alignment of molecules will cause the material to change dimensions. This phenomenon is known as electrostriction. In addition, a permanently-polarized material such as quartz (SiO_2) or barium titanate ($BaTiO_3$) will produce an electric field when the material changes dimensions as a result of an imposed mechanical force. This phenomenon is known as the piezoelectric effect. Additional information on why certain materials produce this effect can be found in the linked presentation material, which was produced by the Valpey Fisher Corporation.

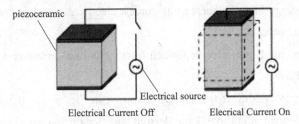

Fig. 6-21 Piezoelectric Transducers

Piezoelectric: The generation of electricity or of electric polarity in ceramic dielectric crystals subjected to mechanical stress, or the generation of stress in such ceramic crystals subjected to an applied voltage.

Magnetostrictive: Deformation of a ferromagnetic material (such as iron and steel) subjected to a magnetic field.

The active element of most acoustic transducers used today is a piezoelectric ceramic, which can be cut in various ways to produce different wave modes. A large piezoelectric ceramic element can be seen in the image of a sectioned low frequency transducer. Preceding the advent of piezoelectric ceramics in the early 1950s, piezoelectric crystals made from quartz crystals and magnetostrictive materials were primarily used. The active element is still sometimes referred to as the crystal by old timers in the NDT field. When piezoelectric ceramics were introduced, they soon became the dominant material for transducers due to their good piezoelectric properties and their ease of manufacture into a variety of shapes and sizes. They also operate at low voltage and are usable up to about 300 ℃. The first piezoceramic in general use was barium titanate, and that was followed during the 1960s by lead zirconate titanate compositions, which are now the most commonly employed ceramic for making transducers. New materials such as piezo-polymers and composites are also being used in some applications.

The thickness of the active element is determined by the desired frequency of the transducer. A thin wafer element vibrates with a wavelength that is twice its thickness. Therefore, piezoelectric crystals are cut to a thickness that is 1/2 the desired radiated wavelength. The higher the frequency of the transducer, the thinner the active element. The primary reason that high frequency contact transducers are not produced is because the element is very thin and too fragile.

The transducer is a very important part of the ultrasonic instrumentation system. As discussed on the previous page, the transducer incorporates a piezoelectric element, which converts electrical signals into mechanical vibrations (transmit mode) and mechanical vibrations into electrical signals (receive mode) . Many factors, including material, mechanical and electrical construction, and the external mechanical and electrical load conditions, influence the behavior of a transducer. Mechanical construction includes parameters such as the radiation surface area, mechanical damping, housing, connector type and other variables of physical construction. As of this writing, transducer manufacturers are hard pressed when constructing two transducers that have identical performance characteristics.

Ultrasonic: Sound frequencies that are higher that detectable with the human ear. A device that converts one form of energy into another. In ultrasonics, electrical energy is converted to mechanical (sound) energy and visa versa.

Piezoelectric Element: Electricity produced by mechanical pressure on a crystal with low symmetry atomic structure.

A cut away of a typical contact transducer is shown in Fig. 6–22. It was previously learned that the piezoelectric element is cut to 1/2 the desired wavelength. To get as much energy out of the transducer as possible, an impedance matching is placed between the active element and the face of the transducer. Optimal impedance matching is achieved by sizing the matching layer so that its thickness is 1/4 of the desired wavelength (Fig. 6–23). This keeps waves that were reflected within the matching layer in phase when they exit the layer (as illustrated in the image to the right). For contact transducers, the matching layer is made from a material that has an acoustical impedance between the active element and steel. Immersion transducers have a matching layer with an acoustical impedance between the active element and water. Contact transducers also incorporate a wear plate to protect the matching layer and active element from scratching.

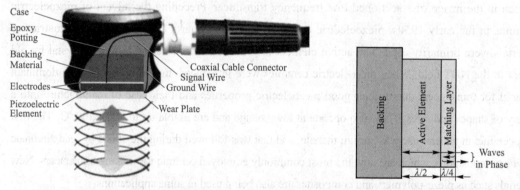

Fig. 6–22 Ultrasonic Transducers   Fig. 6–23 The Active Element and the Backing Material

The backing material supporting the crystal has a great influence on the damping characteristics of a transducer. Using a backing material with an impedance similar to that of the active element will produce the most effective damping. Such a transducer will have a wider bandwidth resulting in higher sensitivity. As the mismatch in impedance between the active element and the backing material increases, material penetration increases but transducer sensitivity is reduced.

Transducer Efficiency, Bandwidth and Frequency

Some transducers are specially fabricated to be more efficient transmitters and others to be more efficient receivers. A transducer that performs well in one application will not always produce the desired results in a different application. For example, sensitivity to small defects is proportional to the product of the efficiency of the transducer as a transmitter and a receiver. Resolution, the ability to locate defects near the surface or in close proximity in the material, requires a highly damped transducer.

It is also important to understand the concept of bandwidth, or range of frequencies, associated with a transducer. The frequency noted on a transducer is the central or center frequency and depends primarily on the backing material. Highly damped transducers will respond to frequencies above and below the central frequency. The broad frequency range provides a transducer with high resolving power. Less damped transducers will exhibit a narrower frequency range and poorer resolving power, but greater penetration. The central frequency will also define the capabilities of a transducer. Lower frequencies (0.5 MHz-2.25 MHz) provide greater energy and penetration in a material, while high frequency crystals (15.0 MHz-25.0 MHz) provide reduced penetration but greater sensitivity to small discontinuities. High frequency transducers, when used with the proper instrumentation, can improve flaw resolution and thickness measurement capabilities dramatically. Broadband transducers with frequencies up to 150 MHz are commercially available.

Transducers are constructed to withstand some abuse, but they should be handled carefully. Misuse, such as dropping, can cause cracking of the wear plate, element, or the backing material. Damage to a transducer is often noted on the A-scan presentation as an enlargement of the initial pulse.

The sound that emanates from a piezoelectric transducer does not originate from a point, but instead originates from most of the surface of the piezoelectric element. Round transducers are often referred to as piston source transducers because the sound field resembles a cylindrical mass in front of the transducer. The sound field from a typical piezoelectric transducer is shown in Fig. 6–24. The intensity of the sound is indicated by color, with lighter colors indicating higher intensity.

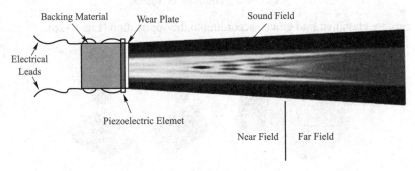

Fig. 6–24 The Sound Field

Since the ultrasound originates from a number of points along the transducer face, the ultrasound intensity along the beam is affected by constructive and destructive wave interference as discussed in a previous page on wave interference. These are sometimes also referred to as diffraction effects. This wave interference leads to extensive fluctuations in the sound intensity near the source and is known as the near field. Because of acoustic variations within a near field, it can be extremely difficult to accurately evaluate flaws in materials when they are positioned within this area.

The pressure waves combine to form a relatively uniform front at the end of the near field. The area beyond the near field where the ultrasonic beam is more uniform is called the far field. In the far field, the beam spreads out in a pattern originating from the center of the transducer. The transition between the near field and the far field occurs at a distance, N, and is sometimes referred to as the "natural focus" of a flat (or unfocused) transducer. The near/far field distance, N, is significant because amplitude variations that characterize the near field change to a smoothly declining amplitude at this point. The area just beyond the near field is where the sound wave is well behaved and at its maximum strength. Therefore, optimal detection results will be obtained when flaws occur in this area.

■ 〖Point 2〗 Transducer Types (Fig. 6–25)

Ultrasonic transducers are manufactured for a variety of applications and can be custom fabricated when necessary. Careful attention must be paid to selecting the proper transducer for the application. A previous section on Acoustic Wavelength and Defect Detection gave a brief overview of factors that affect defect detectability. From this material, we know that it is important to choose transducers that have the desired frequency, bandwidth, and focusing to optimize inspection capability. Most often the transducer is chosen either to enhance the sensitivity or resolution of the system.

Fig. 6–25   Transducer Types

Transducers are classified into groups according to the application (Fig. 6–26).

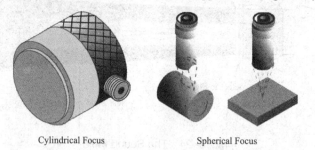

Cylindrical Focus   Spherical Focus

Fig. 6–26   Focus

Contact transducers (Fig. 6–27) are used for direct contact inspections, and are generally hand manipulated. They have elements protected in a rugged casing to withstand sliding contact with a variety of materials. These transducers have an ergonomic design so that they are easy to grip and move along a surface. They often have replaceable wear plates to lengthen their useful life. Coupling materials of water, grease, oils, or commercial materials are used to remove the air gap between the transducer and the component being inspected.

Immersion transducers do not contact the component. These transducers are designed to operate in a liquid environment and all connections are watertight. Immersion transducers usually have an impedance matching layer that helps to get more sound energy into the water, in turn, into the component being inspected. Immersion transducers can be purchased with a planer, cylindrically focused or spherically focused lens. A focused transducer can improve the sensitivity and axial resolution by concentrating the sound energy to a smaller area. Immersion transducers are typically used inside a water tank or as part of a squirter or bubbler system in scanning applications.

Contact transducers are available in a variety of configurations to improve their usefulness for a variety of applications. The flat contact transducer shown above

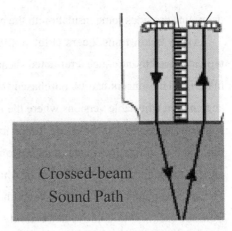

Fig. 6–27   Contact Transducers

is used in normal beam inspections of relatively flat surfaces, and where near surface resolution is not critical. If the surface is curved, a shoe that matches the curvature of the part may need to be added to the face of the transducer. If near surface resolution is important or if an angle beam inspection is needed, one of the special contact transducers described below might be used.

Dual element transducers (Fig. 6–28) contain two independently operated elements in a single housing. One of the elements transmits and the other receives the ultrasonic signal. Active elements can be chosen for their sending and receiving capabilities to provide a transducer with a cleaner signal, and transducers for special applications, such as the inspection of course grained material. Dual element transducers are especially well

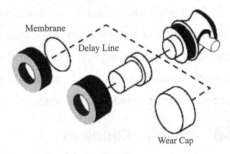

Fig. 6–28   Dual Element Transducers

suited for making measurements in applications where reflectors are very near the transducer since this design eliminates the ring down effect that single-element transducers experience (when single-element transducers are operating in pulse echo mode, the element cannot start receiving reflected signals until the element has stopped ringing from its transmit function) . Dual element transducers are very useful when making thickness measurements of thin materials and when inspecting for near surface defects. The two elements are angled towards each other to create a crossed-beam sound path in the test material.

Delay line transducers provide versatility with a variety of replaceable options. Removable delay line, surface conforming membrane, and protective wear cap options can make a single transducer effective for a wide range of applications. As the name implies, the primary function of a delay line transducer is to introduce a time delay between the generation of the sound wave and the arrival of any reflected waves. This allows the transducer to complete its "sending" function before it starts its "listening" function so that near surface resolution is improved. They are designed for use in applications such as high precision thickness gauging of thin materials and delamination checks in composite materials. They are also useful in high-temperature measurement applications since the delay line provides some insulation to the piezoelectric element from the heat.

Angle beam transducers (Fig. 6-29) and wedges are typically used to introduce a refracted shear wave into the test material. Transducers can be purchased in a variety of fixed angles or in adjustable versions where the user determines the angles of incidence and refraction. In the fixed angle versions, the angle of refraction that is marked on the transducer is only accurate for a particular material, which is usually steel. The angled sound path allows the sound beam to be reflected from the backwall to improve detectability of flaws in and around welded areas. They are also used to generate surface waves for use in detecting defects on the surface of a component.

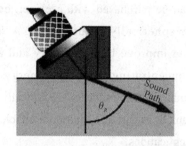

Fig. 6-29 Angle Beam Transducers

Normal incidence shear wave transducers are unique because they allow the introduction of shear waves directly into a test piece without the use of an angle beam wedge. Careful design has enabled manufacturing of transducers with minimal longitudinal wave contamination. The ratio of the longitudinal to shear wave components is generally below -30 dB.

Paint brush transducers are used to scan wide areas. These long and narrow transducers are made up of an array of small crystals that are carefully matched to minimize variations in performance and maintain uniform sensitivity over the entire area of the transducer. Paint brush transducers make it possible to scan a larger area more rapidly for discontinuities. Smaller and more sensitive transducers are often then required to further define the details of a discontinuity.

■ 〖Point 3〗 Couplant

A couplant (Fig. 6-30) is a material (usually liquid) that facilitates the transmission of ultrasonic energy from the transducer into the test specimen. Couplant is generally necessary because the acoustic impedance mismatch between air and solids (i.e. such as the test specimen) is large.

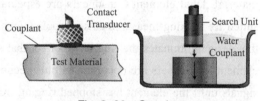

Fig. 6-30 Couplant

Therefore, nearly all of the energy is reflected and very little is transmitted into the test material. The couplant displaces the air and makes it possible to get more sound energy into the test specimen so that a usable ultrasonic signal can be obtained. In contact ultrasonic testing a thin film of oil, glycerin or water is generally used between the transducer and the test surface.

Piezoelectric: the generation of electricity or of electric polarity in ceramic dielectric crystals subjected to mechanical stress, or the generation of stress in such ceramic crystals subjected to an applied voltage.

Magnetostrictive: deformation of a ferromagnetic material (such as iron and steel) subjected to a magnetic field.

When scanning over the part or making precise measurements, an immersion technique is often used. In immersion ultrasonic testing both the transducer and the part are immersed in the couplant, which is typically water. This method of coupling makes it easier to maintain consistent coupling while moving and manipulating the transducer and/or the part.

■ 〖Point 4〗 Pulser-Receivers

Ultrasonic pulser-receivers (Fig. 6-31) are well suited to general purpose ultrasonic testing. Along with appropriate transducers and an oscilloscope, they can be used for flaw detection and thickness gauging in a wide variety of metals, plastics, ceramics, and composites. Ultrasonic pulser-receivers provide a unique, low-cost ultrasonic measurement capability.

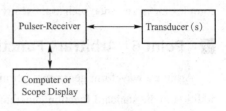

Fig. 6-31　Ultrasonic Pulser-receivers

The pulser section of the instrument generates short, large amplitude electric pulses of controlled energy, which are converted into short ultrasonic pulses when applied to an ultrasonic transducer. Most pulser sections have very low impedance outputs to better drive transducers. Control functions associated with the pulser circuit include:

• Pulse length or damping (The amount of time the pulse is applied to the transducer.)

• Pulse energy (The voltage applied to the transducer. Typical pulser circuits will apply from 100 to 800 Volts to a transducer.)

In the receiver section the voltage signals produced by the transducer, which represent the received ultrasonic pulses, are amplified. The amplified radio frequency (RF) signal is available as an output for display or capture for signal processing. Control functions associated with the receiver circuit include:

• Signal rectification (The RF signal can be viewed as positive half wave, negative half wave or full wave.)

• Filtering to shape and smooth return signals

• Gain, or signal amplification

• Reject control

The pulser-receiver is also used in material characterization work involving sound velocity or attenuation measurements, which can be correlated to material properties such as elastic modulus. In conjunction with a stepless gate and a spectrum analyzer, pulser-receivers are also used to study frequency dependent material properties or to characterize the performance of ultrasonic transducers.

■ 〖Point 5〗 Tone Burst Generators in Research

Resonance: reinforcement and prolongation of a sound or musical tone by reflection or by sympathetic vibration of other bodies.

Superheterodyne Receiver: donating a device or method of radio reception in which beats are produced by superimposing a locally generated radio wave on an incoming wave. In the superheterodyne receiver the intermediate frequency is amplified and demodulated.

Tone burst generators are often used in high power ultrasonic applications. They take low-voltage signals and convert them into high-power pulse trains for the most power-demanding applications. Their purpose is to transmit bursts of acoustic energy into a test piece, receive the resulting signals, and then manipulate and analyze the received signals in various ways. High power radio frequency (RF) burst capability allows researchers to work with difficult, highly attenuative materials or inefficient transducers such as EMATs. A computer interface makes it possible for systems to make high speed complex measurements, such as those involving multiple frequencies.

■ 〖Point 6〗 Arbitrary Function Generators

Arbitrary waveform generators permit the user to design and generate virtually any waveform in addition to the standard function generator signals (i.e. sine wave, square wave, etc.) . Waveforms are generated digitally from a computer's memory, and most instruments allow the downloading of digital waveform files from computers.

Ultrasonic generation pulses must be varied to accommodate different types of ultrasonic transducers. General-purpose highly damped contact transducers are usually excited by a wideband, spike-like pulse provided by many common pulser/receiver units. The lightly damped transducers used in high power generation, for example, require a narrowband tone-burst excitation from a separate generator unit. Sometimes the same transducer will be excited differently, such as in the study of the dispersion of a material's ultrasonic attenuation or to characterize ultrasonic transducers.

In spread spectrum ultrasonics (Fig. 6-32) , encoded sound is generated by an arbitrary waveform generator continuously transmitting coded sound into the part or structure being tested. Instead of receiving echoes, spread spectrum ultrasonics generates an acoustic correlation signature having a one-to-one correspondence with the acoustic state of the part or structure (in its environment) at the instant of measurement. In its simplest embodiment, the acoustic correlation signature is generated by cross correlating an encoding sequence (with suitable cross and auto correlation properties) transmitted into a part (structure) with received signals returning from the part (structure) .

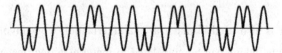

Fig. 6-32 Section of Biphase Modulated Spread Spectrum Ultrasonic Waveform

■ 〖Point 7〗 Electrical Impedance Matching and Termination

When computer systems were first introduced decades ago, they were large, slow-working devices that were incompatible with each other. Today, national and international networking standards have established electronic control protocols that enable different systems to "talk" to each other. The Electronics Industries Associations (EIA) and the Institute of Electrical and Electronics Engineers (IEEE) developed standards that established common terminology and interface requirements, such as EIA RS-232 and IEEE 802.3. If a system designer builds equipment to comply with these standards, the equipment will interface with other systems. But what about analog signals that are used in ultrasonics?

Data Signals: Input Versus Output

Consider the signal going to and from ultrasonic transducers. When you transmit data through a cable, the requirement usually simplifies into comparing what goes in one end with what comes out the other. High frequency pulses degrade or deteriorate when they are passed through any cable. Both the height of the pulse (magnitude) and the shape of the pulse (waveform) change dramatically, and the amount of change depends on the data rate, transmission distance and the cable's electrical characteristics. Sometimes a marginal electrical cable may perform adequately if used in only short lengths, but the same cable with the same data in long lengths will fail. This is why system designers and industry standards specify precise cable criteria.

Cable Electrical Characteristics

Impedance: The quantity that measures the opposition of a circuit to the passage of a current and therefore determines the amplitude of the current.

Attenuation: a loss of intensity suffered by sound, radiation, etc., as it passes through a medium.

Shielding: a barrier surrounding a region to exclude it from the influence of an energy field.

Capacitance: the property of a conductor or system of conductors that describes its ability to store electric charge.

The most important characteristics in an electronic cable are impedance, attenuation, shielding, and capacitance. In this page, we can only review these characteristics very generally, however, we will discuss capacitance in more detail.

Impedance (Ohms) represents the total resistance that the cable presents to the electrical current passing through it. At low frequencies the impedance is largely a function of the conductor size, but at high frequencies conductor size, insulation material, and insulation thickness all affect the cable's impedance. Matching impedance is very important. If the system is designed to be 100 ohms, then the cable should match that impedance, otherwise error-producing reflections are created.

Attenuation is measured in decibels per unit length (dB/m), and provides an indication of the signal loss as it travels through the cable. Attenuation is very dependent on signal frequency. A cable that works very well with low frequency data may do very poorly at higher data rates. Cables with lower attenuation are better.

Shielding is normally specified as a cable construction detail. For example, the cable may be unshielded, contain shielded pairs, have an overall aluminum/mylar tape and drain wire, or have a double shield. Cable shields usually have two functions: to act as a barrier to keep external signals from getting in and internal signals from getting out, and to be a part of the electrical circuit. Shielding effectiveness is very complex to measure and depends on the data frequency within the cable and the precise shield design. A shield may be very effective in one frequency range, but a different frequency may require a completely different design. System designers often test complete cable assemblies or connected systems for shielding effectiveness.

Recommendations:

• Observe manufacturer's recommended practices for cable impedance, cable length, impedance matching, and any requirements for termination in characteristic impedance.

• If possible, use the same cables and cable dressing for all inspections.

Capacitance in a cable is usually measured as picofarads per foot (pf/m) . It indicates how much charge the cable can store within itself. If a voltage signal is being transmitted by a twisted pair, the insulation of the individual wires becomes charged by the voltage within the circuit. Since it takes a certain amount of time for the cable to reach its charged level, this slows down and interferes with the signal being transmitted. Digital data pulses are a string of voltage variations that are represented by square waves. A cable with a high capacitance slows down these signals so that they come out of the cable looking more like "saw-teeth", rather than square waves. The lower the capacitance of the cable, the better it performs with high speed data.

〖Point 8〗 Data Presentation

Ultrasonic data can be collected and displayed in a number of different formats. The three most common formats are knowing in the NDT world as A-scan, B-scan and C-scan presentations. Each presentation mode provides a different way of looking at and evaluating the region of material being inspected. Modern computerized ultrasonic scanning systems can display data in all three presentation forms simultaneously.

A-scan Presentation

The A-scan presentation (Fig. 6–33) displays the amount of received ultrasonic energy as a function of time. The relative amount of received energy is plotted along the vertical axis and the elapsed time (which may be related to the sound energy travel time within the material) is displayed along the horizontal axis. Most instruments with an A-scan display (Fig. 6–34) allow the signal to be displayed in its natural radio frequency form (RF), as a fully rectified RF signal, or as either the positive or negative half of the RF signal. In the A-scan presentation, relative discontinuity size can be estimated by comparing the signal amplitude obtained from an unknown reflector to that from a known reflector. Reflector depth can be determined by the position of the signal on the horizontal sweep.

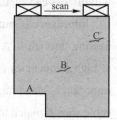

Fig. 6–33 A-scan Presentation

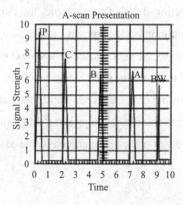
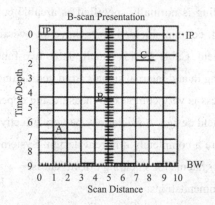

Fig. 6–34 Display

In the illustration of the A-scan presentation to the right, the initial pulse generated by the transducer is represented by the signal IP, which is near time zero. As the transducer is scanned along the surface of the part, four other signals are likely to appear at different times on the screen. When the transducer is in its far left position, only the IP signal and signal A, the sound energy reflecting from surface A, will be seen on the trace. As the transducer is scanned to the right, a signal from the backwall BW will appear later in time, showing that the sound has traveled farther to reach this surface. When the transducer is over flaw B, signal B will appear at a point on the time scale that is approximately halfway between the IP signal and the BW signal. Since the IP signal corresponds to the front surface of the material, this indicates that flaw B is about halfway between the front and back surfaces of the sample. When the transducer is moved over flaw C, signal C will appear earlier in time since the sound travel path is shorter and signal B will disappear since sound will no longer be reflecting from it.

B-scan Presentation

The B-scan presentations are a profile (cross-sectional) view of the test specimen. In the B-scan, the time-of-flight (travel time) of the sound energy is displayed along the vertical axis and the linear position of the transducer is displayed along the horizontal axis. From the B-scan, the depth of the reflector and its approximate linear dimensions in the scan direction can be determined. The B-scan is typically produced by establishing a trigger gate on the A-scan. Whenever the signal intensity is great enough to trigger the gate, a point is produced on the B-scan. The gate is triggered by the sound reflecting from the backwall of the specimen and by smaller reflectors within the material. In the B-scan image above, line A is produced as the transducer is scanned over the reduced thickness portion of the specimen. When the transducer moves to the right of this section, the backwall line BW is produced. When the transducer is over flaws B and C, lines that are similar to the length of the flaws and at similar depths within the material are drawn on the B-scan. It should be noted that a limitation to this display technique is that reflectors may be masked by larger reflectors near the surface.

C-scan Presentation

The C-scan presentation (Fig. 6–35) provides a plan-type view of the location and size of test specimen features. The plane of the image is parallel to the scan pattern of the transducer. C-scan presentations are produced with an automated data acquisition system, such as a computer controlled immersion scanning system. Typically, a data collection gate is established on the A-scan and the amplitude or the time-of-flight of the signal is recorded at regular intervals as the transducer is

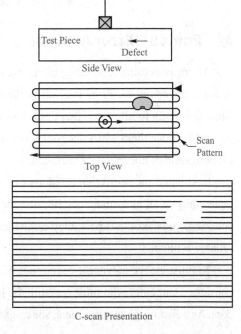

Fig. 6–35 The C-scan Presentation

scanned over the test piece. The relative signal amplitude or the time-of-flight is displayed as a shade of gray or a color for each of the positions where data was recorded. The C-scan presentation provides an image of the features that reflect and scatter the sound within and on the surfaces of the test piece.

High resolution scans can produce very detailed images. Fig. 6–36 are two ultrasonic C-scan images of a US quarter. Both images were produced using a pulse-echo technique with the transducer scanned over the head side in an immersion scanning system. For the C-scan image on the left, the gate was setup to capture the amplitude of the sound reflecting from the front surface of the quarter. Light areas in the image indicate areas that reflected a greater amount of energy back to the transducer. In the C-scan image on the right, the gate was moved to record the intensity of the sound reflecting from the back surface of the coin. The details on the back surface are clearly visible but front surface features are also still visible since the sound energy is affected by these features as it travels through the front surface of the coin.

Fig. 6–36 C-scan of the Coin

〖Point 9〗 Error Analysis

All measurements, including ultrasonic measurements, however careful and scientific, are subject to some uncertainties. Error analysis is the study and evaluation of these uncertainties; its two main functions being to allow the practitioner to estimate how large the uncertainties are and to help him or her to reduce them when necessary. Because ultrasonics depends on measurements, evaluation and minimization of uncertainties is crucial.

In science the word "error" does not mean "mistake" or "blunder" but rather the inevitable uncertainty of all measurements. Because they cannot be avoided, errors in this context are not, strictly speaking, "mistakes". At best, they can be made as small as reasonably possible, and their size can be reliably estimated.

To illustrate the inevitable occurrence of uncertainties surrounding attempts at measurement, let us consider a carpenter who must measure the height of a doorway to an X-ray vault in order to install a door. As a first rough measurement, she might simply look at the doorway and estimate that it is 210 cm high. This crude "measurement" is certainly subject to uncertainty. If pressed, the carpenter might express this uncertainty by admitting that the height could be as little as 205 cm or as much as 215 cm.

If she wanted a more accurate measurement, she would use a tape measure, and she might find that the height is 211.3 cm. This measurement is certainly more precise than her original estimate, but it is obviously still subject to some uncertainty, since it is inconceivable that she could know the height to be exactly 211.300 0 rather than 211.300 1 cm, for example.

There are many reasons for this remaining uncertainty. Some of these causes of uncertainty could be removed if enough care were taken. For example, one source of uncertainty might be that poor lighting is making it difficult to read the tape; this could be corrected by improved lighting.

On the other hand, some sources of uncertainty are intrinsic to the process of measurement and can never be entirely removed. For instance, let us suppose the carpenter's tape is graduated in half-centimeters. The top of the door will probably not coincide precisely with one of the half-centimeter marks, and if it does not, then the carpenter must estimate just where the top lies between two marks. Even if the top happens to coincide with one of the marks, the mark itself is perhaps a millimeter wide, so she must estimate just where the top lies within the mark. In either case, the carpenter ultimately must estimate where the top of the door lies relative to the markings on her tape, and this necessity causes some uncertainty in her answer.

By buying a better tape with closer and finer markings, the carpenter can reduce her uncertainty, but she cannot eliminate it entirely. If she becomes obsessively determined to find the height of the door with the greatest precision that is technically possible, she could buy an expensive laser interferometer. But even the precision of an interferometer is limited to distances on the order of the wavelength of light (about 0.000 005 meters). Although she would now be able to measure the height with fantastic precision, she still would not know the height of the doorway exactly.

Furthermore, as the carpenter strives for greater precision, she will encounter an important problem of principle. She will certainly find that the height is different in different places. Even in one place, she will find that the height varies if the temperature and humidity vary, or even if she accidentally rubs off a thin layer of dirt. In other words, she will find that there is no such thing as one exact height of the doorway. This kind of problem is called a "problem of definition" (the height of the door is not well-defined and plays an important role in many scientific measurements).

Our carpenter's experiences illustrate what is found to be generally true. No physical quantity (a thickness, time between pulse-echoes, a transducer position, etc.) can be measured with complete certainty. With care we may be able to reduce the uncertainties until they are extremely small, but to eliminate them entirely is impossible.

In everyday measurements we do not usually bother to discuss uncertainties. Sometimes the uncertainties are simply not interesting. If we say that the distance between home and school is 3 miles, it does not matter (for most purposes) whether this means "somewhere between 2.5 and 3.5 miles" or "somewhere between 2.99 and 3.01 miles". Often the uncertainties are important, but can be allowed for instinctively and without explicit consideration. When our carpenter comes to fit her door, she must know its height with an uncertainty that is less than 1 mm or so. However, as long as the uncertainty is this small, the door will (for all practical purposes) be a perfect fit, X-rays will not leak out, and her concern with error analysis will come to an end.

Put into Practice

1. Translate Fig. 6–37 and explain the type of probe and what the efficacy of each section is.

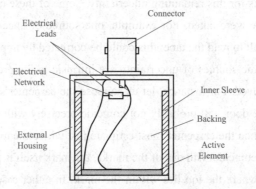

Fig. 6–37   Ultrasonic Testing Probe

2. Translate Fig. 6–38 and explain the type of probe and what the efficacy of each section is.

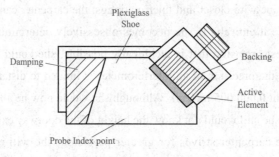

Fig. 6–38   Ultrasonic Testing Probe

3. Fig. 6–39 shows the difference between the probe type and the detection principle.

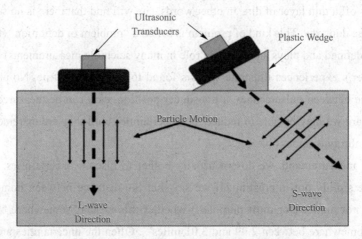

Fig. 6–39   Longitudinal and Shear Waves

238

6.4 Measurement Techniques

■ 〚Point 1〛 Angle Beams I (Fig. 6-40)

Refraction: the change of direction suffered by wavefront as it passes obliquely from one medium to another in which its speed of propagation is altered.

Angle Beam Transducers and wedges are typically used to introduce a refracted shear wave into the test material. An angled sound path allows the sound beam to come

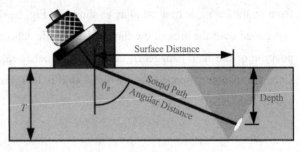

Fig. 6-40 Angle Beams I

in from the side, thereby improving detectability of flaws in and around welded areas.

θ_R=Angle of Refraction
T=Material Thickness
Surface Distance=$\sin\theta_R \times$ Sound Path
Depth (1st Leg) =$\cos\theta_R \times$ Sound Path

■ 〚Point 2〛 Angle Beams II (Fig. 6-41)

Angle Beam Transducers and wedges are typically used to introduce a refracted shear wave into the test material. The geometry of the sample below allows the sound beam to be reflected from the back wall to improve detectability of flaws in and around welded areas.

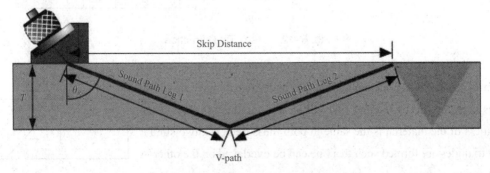

Fig. 6-41 Angle Beams II

θ_R=Refracted Angle
T=Material Thickness
Skip=Distance=$2T \times \tan\theta_R$

$\text{Leg} = \dfrac{T}{\cos\theta_R}$

$\text{V-Path} = \dfrac{2T}{\cos\theta_R}$

〖Point 3〗 Crack Tip Diffraction

Diffraction: the spreading or bending of waves as they pass through an aperture (transducer) or round the edge of a barrier.

When the geometry of the part is relatively uncomplicated and the orientation of a flaw is well known, the length (*a*) of a crack can be determined by a technique known as tip diffraction. One common application of the tip diffraction technique is to determine the length of a crack originating from on the backside of a flat plate as shown in Fig. 6–42. In this case, when an angle beam transducer is scanned over the area of the flaw, the principle echo comes from the base of the crack to locate the position of the flaw [Image(a)] . A second, much weaker echo comes from the tip of the crack and since the distance traveled by the ultrasound is less, the second signal appears earlier in time on the scope [Image(b)].

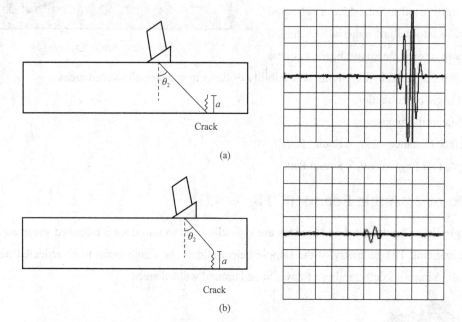

Fig. 6–42 Crack Tip Diffraction

Crack height (*a*) (Fig. 6–43) is a function of the ultrasound velocity (*v*) in the material, the incident angle (θ_2) and the difference in arrival times between the two signal (d*t*) . Since the incident angle and the thickness of the material is the same in both measurements, two similar right triangles are formed such that one can be overlayed on the other. A third similar right triangle is made, which is comprised on the crack, the length d*t* and the angle θ_2. The variable d*t* is really the difference in time but can easily be converted to a distance by dividing the time in half (to get the one-way travel time) and multiplying this value by the velocity of the sound in the material. Using trigonometry an equation for estimating crack height from these variables can be derived .

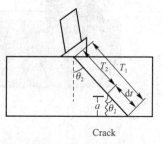

Fig. 6–43 Crack Height

〔Point 4〕 Precision Velocity Measurements

Signal Processing Techniques

Signal processing involves techniques that improve our understanding of information contained in received ultrasonic data. Normally, when a signal is measured with an oscilloscope, it is viewed in the time domain (vertical axis is amplitude or voltage and the horizontal axis is time). For many signals, this is the most logical and intuitive way to view them. Simple signal processing often involves the use of gates to isolate the signal of interest or frequency filters to smooth or reject unwanted frequencies.

When the frequency content of the signal is of interest, it makes sense to view the signal graph in the frequency domain. In the frequency domain, the vertical axis is still voltage but the horizontal axis is frequency.

Time domain and frequency domain magnitude are shown in Fig. 6-44.

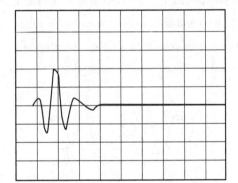 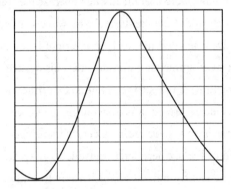

Fig. 6-44 Time Domain (left) and Frequency Domain Magnitude (right)

The frequency domain display shows how much of the signal's energy is present as a function of frequency. For a simple signal such as a sine wave, the frequency domain representation does not usually show us much additional information. However, with more complex signals, such as the response of a broad bandwidth transducer, the frequency domain gives a more useful view of the signal.

Fourier theory says that any complex periodic waveform can be decomposed into a set of sinusoids with different amplitudes, frequencies and phases. The process of doing this is called Fourier Analysis, and the result is a set of amplitudes, phases, and frequencies for each of the sinusoids that makes up the complex waveform. Adding these sinusoids together again will reproduce exactly the original waveform. A plot of the frequency or phase of a sinusoid against amplitude is called a spectrum.

The following Fourier Java applet, adapted with permission of Stanford University, allows the user to manipulate discrete time domain or frequency domain components and see the relationships between signals in time and frequency domains.

The top row (light blue color) represents the real and imaginary parts of the time domain. Normally the imaginary part of the time domain signal is identically zero.

The middle row (peach color) represents the the real and imaginary parts of the frequency domain.

The bottom row (light green color) represents the magnitude (amplitude) and phase of the frequency domain signal. Magnitude is the square root of the sum of the squares of the real and imaginary components. Phase is the angular relationship of the real and imaginary components. Ultrasonic transducer manufactures often provide plots of both time domain and frequency domain (magnitude) signals characteristic of each transducer. Use this applet to explore the relationship between time and frequency domains.

〖Point 5〗 Flaw Reconstruction Techniques

In non-destructive evaluation of structural material defects, the size, shape, and orientation are important flaw parameters in structural integrity assessment. To illustrate flaw reconstruction, a multiviewing ultrasonic transducer system is shown in Fig. 6–45. A single probe moved sequentially to achieve different perspectives would work equally as well. The apparatus and the signal-processing algorithms were specifically designed at the Center for non-destructive evaluation to make use of the theoretical developments in elastic wave scattering in the long and intermediate wavelength regime.

Depicted schematically at the right is the multiprobe system consisting of a sparse array of seven unfocused immersion transducers. This system can be used to "focus" onto a target flaw in a solid by refraction at the surface. The six perimeter transducers are equally spaced on a 5.08 cm diameter ring, surrounding a center transducer. Each of the six perimeter transducers may be independently moved along its axis to allow an equalization of the propagation time for any pitch-catch or pulse-echo combinations. The system currently uses 0.25 in diameter transducers with a nominal center frequency of 10 MHz and a bandwidth extending from approximately 2 to 16 MHz. The axis of the aperture cone of the transducer assembly normally remains vertical and perpendicular to the part surface.

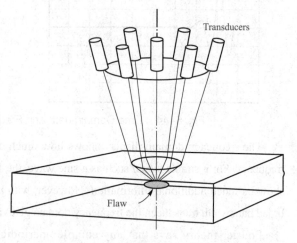

Fig. 6–45 Flaw Reconstruction Techniques

The flaw reconstruction algorithm normally makes use of 13 or 19 backscatter waveforms acquired in a conical pattern within the aperture. The data-acquisition and signal-processing protocol has four basic steps.

(1) Step one involves the experimental setup, the location and focusing on a target flaw, and acquisition (in a predetermined pattern) of pitch-catch and pulse-echo backscatter waveforms.

(2) Step two employs a measurement model to correct the backscatter waveforms for effects of attenuation, diffraction, interface losses, and transducer characteristics, thus resulting in absolute scattering amplitudes.

(3) Step three employs a one-dimensional inverse Born approximation to extract a tangent plane

to centroid radius estimate for each of the scattering amplitudes.

(4) In step four the radius estimates and their corresponding look angles are used in a regression analysis program to determine the six ellipsoidal parameters, three semiaxes, and three Euler angles, defining an ellipsoid which best fits the data.

The inverse Born approximation sizes the flaw by computing the characteristic function of the flaw (defined as unity inside the flaw and zero outside the flaw) as a Fourier transform of the ultrasonic scattering amplitude. The one-dimensional inverse Born algorithm treats scattering data in each interrogation direction independently and has been shown to yield the size of ellipsoidal flaws (both voids and inclusions) in terms of the distance from the center of the flaw to the wavefront that is tangent to the front surface of the flaw. Using the multiprobe ultrasonic system, the 1-D Inverse Born technique is used to reconstruct voids and inclusions that can be reasonably approximated by an equivalent ellipsoid. So far, the investigation has been confined to convex flaws with a center of inversion symmetry. The angular scan method described in this paper is capable of locating the bisecting symmetry planes of a flaw. The utility of the multiprobe system is, therefore, expanded since two-dimensional elliptic reconstruction may now be made for the central slice. Additionally, the multiprobe system is well suited for the 3-D flaw reconstruction technique using 2-D slices.

The model-based reconstruction method has been previously applied to voids and incursion flaws in solids. Since the least-squares regression analysis leading to the "best fit" ellipsoid is based on the tangent plane to centroid distances for the interrogation directions confined within a finite aperture. The success of reconstruction depends on the extent of the flaw surface "illuminated" by the various viewing directions. The extent of coverage of the flaw surface by the tangent plane is a function of the aperture size, flaw shape, and the flaw orientation. For example, a prolate spheroidal flaw with a large aspect ratio oriented along the axis of the aperture cone will only have one tip illuminated (i.e. covered by the tangent planes) and afford a low reconstruction reliability. For the same reason, orientation of the flaw also has a strong effect on the reconstruction accuracy.

The diagram on the right shows the difference in surface coverage of a tilted flaw and an untilted flaw subjected to the same insonification aperture. Both the experimental and simulation studies of the aperture effect reported before were conducted for oblate and prolate spheroids oriented essentially symmetrically with respect to the part surface and hence the aperture cone. From a flaw reconstruction standpoint, an oblate spheroid with its axis of rotational symmetry perpendicular to the part surface represents a high leverage situation. Likewise, a prolate spheroid with its symmetry axis parallel to the part surface also affords an easier reconstruction than a tilted prolate spheroid. In this CNDE project, we studied effects of flaw orientation on the reconstruction and derived a new data-acquisition approach that will improve reliability of the new reconstruction of arbitrarily oriented flaws.

The orientation of a flaw affects reconstruction results in the following ways.

(1) For a given finite aperture, a change in flaw orientation will change the insonified surface area and hence change the "leverage" for reconstruction.

(2) The scattering signal amplitude and the signal/noise ratio for any given interrogation direction

depends on the flaw orientation.

(3) Interference effects, such as those due to tip diffraction phenomena or flash points may be present at certain orientations. Of course, interdependencies exist in these effects, but for the sake of convenience they are discussed separately in the following.

■ 〔Point 6〕 Calibration Methods

Calibration (Fig. 6-46) refers to the act of evaluating and adjusting the precision and accuracy of measurement equipment. In ultrasonic testing, several forms of calibration must occur. First, the electronics of the equipment must be calibrated to ensure that they are performing as designed. This operation is usually performed by the equipment manufacturer and will not

Fig. 6-46 Calibration

be discussed further in this material. It is also usually necessary for the operator to perform a "user calibration" of the equipment. This user calibration is necessary because most ultrasonic equipment can be reconfigured for use in a large variety of applications. The user must "calibrate" the system, which includes the equipment settings, the transducer and the test setup, to validate that the desired level of precision and accuracy are achieved. The term calibration standard is usually only used when an absolute value is measured and in many cases, the standards are traceable back to standards at the National Institute for Standards and Technology.

In ultrasonic testing, there is also a need for reference standards. Reference standards are used to establish a general level of consistency in measurements and to help interpret and quantify the information contained in the received signal. Reference standards are used to validate that the equipment and the setup provide similar results from one day to the next and that similar results are produced by different systems. Reference standards also help the inspector to estimate the size of flaws. In a pulse-echo type setup, signal strength depends on both the size of the flaw and the distance between the flaw and the transducer. The inspector can use a reference standard with an artificially induced flaw of known size and at approximately the same distance away for the transducer to produce a signal. By comparing the signal from the reference standard to that received from the actual flaw, the inspector can estimate the flaw size.

This section will discuss some of the more common calibration and reference specimen that are used in ultrasonic inspection. Some of these specimens are shown in the figure above. Be aware that there are other standards available and that specially designed standards may be required for many applications. The information provided here is intended to serve a general introduction to the standards and not to be instruction on the proper use of the standards.

Introduction to the Common Standards

Calibration and reference standards for ultrasonic testing come in many shapes and sizes. The type of standard used is dependent on the NDE application and the form and shape of the object being evaluated. The material of the reference standard should be the same as the material being inspected and the artificially induced flaw should closely resemble that of the actual flaw. This second requirement is a major limitation of most standard reference samples. Most use drilled holes and notches that do not closely represent real flaws. In most cases the artificially induced defects in reference standards are better reflectors of sound energy (due to their flatter and smoother surfaces) and produce indications that are larger than those that a similar sized flaw would produce. Producing more "realistic" defects is cost prohibitive in most cases and, therefore, the inspector can only make an estimate of the flaw size. Computer programs that allow the inspector to create computer simulated models of the part and flaw may one day lessen this limitation.

The standard shown in Fig. 6–47 is commonly known in the US as an IIW type reference block. IIW is an acronym for the International Institute of Welding. It is referred to as an IIW "type" reference block because it was patterned after the "true" IIW block but does not conform to IIW requirements in IIS/IIW-23-59. "True" IIW

Fig. 6–47 The IIW Type Calibration Block

blocks are only made out of steel (to be precise, killed, open hearth or electric furnace, low-carbon steel in the normalized condition with a grain size of Mc Quaid-Ehn $^{\#}8$) where IIW "type" blocks can be commercially obtained in a selection of materials. The dimensions of "true" IIW blocks are in metric units while IIW "type" blocks usually have English units. IIW "type" blocks may also include additional calibration and references features such as notches, circular groves, and scales that are not specified by IIW. There are two full-sized and a mini version of the IIW type blocks. The Mini version is about one-half the size of the full-sized block and weighs only about one-fourth as much. The IIW type US-1 block was derived the basic "true" IIW block . The IIW type US-2 block was developed for US Air Force application. A Mini version also exists.

IIW type blocks are used to calibrate instruments for both angle beam and normal incident inspections. Some of their uses include setting metal-distance and sensitivity settings, determining the sound exit point and refracted angle of angle beam transducers, and evaluating depth resolution of normal beam inspection setups. Instructions on using the IIW type blocks can be found in the annex of American Society for Testing and Materials Standard E164, Standard Practice for Ultrasonic Contact Examination of Weldments.

The miniature angle-beam is (Fig. 6–48) a calibration block that was designed for the US Air Force for use in the field for instrument calibration. The block is much smaller and lighter than the IIW block but performs many of the same functions. The miniature angle-beam block can

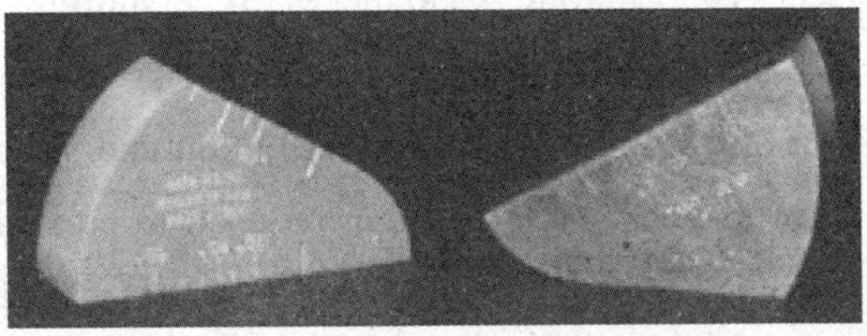

Fig. 6–48 The Miniature Angle–beam or ROMPAS Calibration Block

be used to check the beam angle and exit point of the transducer. The block can also be used to make metal-distance and sensitivity calibrations for both angle and normal-beam inspection setups.

A block that closely resembles the miniature angle-beam block and is used in a similar way is the DSC AWS Block(Fig. 6–49). This block is used to determine the beam exit point and refracted angle of angle-beam transducers and to calibrate distance and set the sensitivity for both normal and angle beam inspection setups. Instructions on using the DSC block can be found in the annex of American Society for Testing and Materials Standard E164, Standard Practice for Ultrasonic Contact Examination of Weldments.

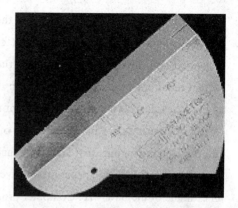

Fig. 6–49 AWS Shear Wave Distance/Sensitivity Calibration (DSC) Block

The DC AWS Block (Fig. 6–50) is a metal path distance and beam exit point calibration standard that conforms to the requirements of the American Welding Society (AWS) and the American Association of State Highway and Transportation Officials (AASHTO) . Instructions on using the DC block can be found in the annex of American Society for Testing and Materials Standard E164, Standard Practice for Ultrasonic Contact Examination of Weldments.

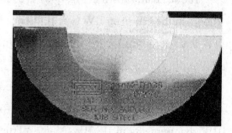

Fig. 6–50 AWS Shear Wave Distance Calibration (DC) Block

The RC Block (Fig. 6–51) is used to determine the resolution of angle beam transducers per the requirements of AWS and AASHTO. Engraved Index markers are provided for 45, 60, and 70 degree refracted angle beams.

Fig. 6–51 AWS Resolution Calibration (RC) Block

The 30 FBH resolution reference block (Fig. 6–52) is used to evaluate the near-surface resolution and flaw size/depth sensitivity of a normal-beam setup. The block contains number 3 (3/64"), 5 (5/64") and 8 (8/64") ASTM flat bottom holes at ten metal-distances ranging from 0.050 inches (1.27 mm) to 1.250 inches (31.75 mm) .

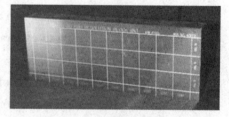

Fig. 6–52 30 FBH Resolution Reference Block

The miniature resolution block (Fig. 6–53) is used to evaluate the near-surface resolution and sensitivity of a normal-beam setup. It can be used to calibrate high-resolution thickness gages over the range of 0.015 inches (0.381 mm) to 0.125 inches (3.175 mm) .

Step and tapered calibration wedges (Fig. 6–54) come

Fig. 6–53 Miniature Resolution Block

in a large variety of sizes and configurations. Step wedges are typically manufactured with four or five steps but custom wedge can be obtained with any number of steps. Tapered wedges have a constant taper over the desired thickness range.

Fig. 6–54   Step and Tapered Calibration Wedges

The DS test block (Fig. 6–55) is a calibration standard used to check the horizontal linearity and the dB accuracy per requirements of AWS and AASHTO.

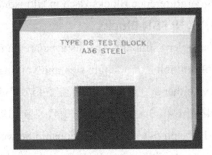

Fig. 6–55   Distance/Sensitivity (DS) Block

Distance/area-amplitude correction blocks typically are purchased as a ten-block set, as shown in Fig. 6–56. Aluminum sets are manufactured per the requirements of ASTM E127 and steel sets per ASTM E428. Sets can also be purchased in titanium. Each block contains a single flat-bottomed, plugged hole. The hole sizes and metal path distances are as follows:

- 3/64" at 3"
- 5/64" at 1/8", 1/4", 1/2", 3/4", 11/2", 3" and 6"
- 8/64" at 3" and 6"

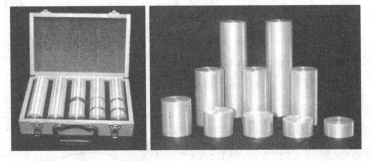

Fig. 6–56   Distance/Area–amplitude Blocks

Sets are commonly sold in 4340 Vacuum melt Steel, 7075-T6 Aluminum, and Type 304 Corrosion Resistant Steel. Aluminum blocks are fabricated per the requirements of ASTM E127, Standard Practice for Fabricating and Checking Aluminum Alloy Ultrasonic Standard Reference Blocks. Steel blocks are fabricated per the requirements of ASTM E428, Standard Practice for Fabrication and Control of Steel Reference Blocks Used in Ultrasonic Inspection.

Area-amplitude Blocks

Area-amplitude blocks are also usually purchased in an eight-block set and look very similar to Distance/Area-Amplitude Blocks. However, area-amplitude blocks have a constant 3-inch metal path distance and the hole sizes are varied from 1/64" to 8/64" in 1/64" steps. The blocks are used to determine the relationship between flaw size and signal amplitude by comparing signal responses for the different sized holes. Sets are commonly sold in 4340 Vacuum melt Steel, 7075-T6 Aluminum, and Type 304 Corrosion Resistant Steel. Aluminum blocks are fabricated per the requirements of ASTM E127, Standard Practice for Fabricating and Checking Aluminum Alloy Ultrasonic Standard Reference Blocks. Steel blocks are fabricated per the requirements of ASTM E428, Standard Practice for Fabrication and Control of Steel Reference Blocks Used in Ultrasonic Inspection.

Distance-amplitude #3, #5, #8 FBH Blocks

Distance-amplitude blocks also very similar to the distance/area-amplitude blocks pictured above. Nineteen block sets with flat-bottom holes of a single size and varying metal path distances are also commercially available. Sets have either a #3 (3/64") FBH, a #5 (5/64") FBH, or a #8 (8/64") FBH. The metal path distances are 1/16", 1/8", 1/4", 3/8", 1/2", 5/8", 3/4", 7/8", 1", 1-1/4", 1-3/4", 2-1/4", 2-3/4", 3-14", 3-3/4", 4-1/4", 4-3/4", 5-1/4" and 5-3/4". The relationship between the metal path distance and the signal amplitude is determined by comparing signals from same size flaws at different depth. Sets are commonly sold in 4340 Vacuum melt Steel, 7075-T6 Aluminum and Type 304 Corrosion Resistant Steel. Aluminum blocks are fabricated per the requirements of ASTM E127, Standard Practice for Fabricating and Checking Aluminum Alloy Ultrasonic Standard Reference Blocks. Steel blocks are fabricated per the requirements of ASTM E428, Standard Practice for Fabrication and Control of Steel Reference Blocks Used in Ultrasonic Inspection.

■ 〖Point 7〗 Distance Amplitude Correction (DAC)

Acoustic signals from the same reflecting surface will have different amplitudes at different distances from the transducer. Distance amplitude correction (DAC) (Fig. 6–57) provides a means of establishing a graphic "reference level sensitivity" as a function of sweep distance on the A-scan display. The use of DAC allows signals reflected from similar discontinuities to be evaluated where signal attenuation as a function of depth has been correlated. Most often DAC will allow for loss in amplitude over material depth (time), graphically on the A-scan

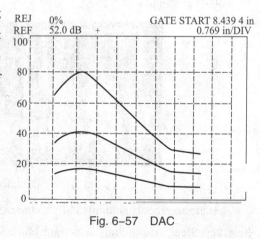

Fig. 6–57 DAC

display but can also be done electronically by certain instruments. Because near field length and beam spread vary according to transducer size and frequency, and materials vary in attenuation and velocity, a DAC curve must be established for each different situation. DAC may be employed in both longitudinal and shear modes of operation as well as either contact or immersion inspection techniques.

A distance amplitude correction curve is constructed from the peak amplitude responses from reflectors of equal area at different distances in the same material. A-scan echoes are displayed at their non-electronically compensated height and the peak amplitude of each signal is marked on the flaw detector screen, preferably, on a transparent plastic sheet attached to the screen. Reference standards which incorporate side drilled holes (SDH), flat bottom holes (FBH), or notches whereby the reflectors are located at varying depths are commonly used. It is important to recognize that regardless of the type of reflector used, the size and shape of the reflector must be constant. Commercially available reference standards for constructing DAC include ASTM Distance/Area-amplitude and ASTM E1158 Distance Amplitude blocks, NAVSHIPS Test block and ASME Basic Calibration Blocks.

Put into Practice

1. Fig. 6–58 explains the principle of defect detection.

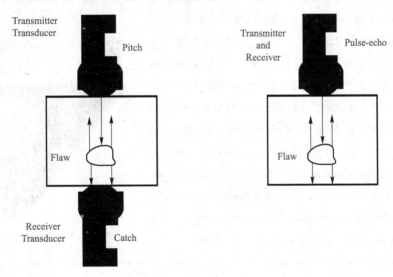

Fig. 6–58   Defect Detection

2. Translate the various displays in Fig. 6–59.

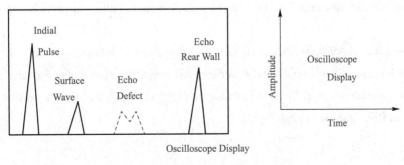

Fig. 6–59   Displays

6.5　Some Applications

■〔Point 1〕Rail Inspection

One of the major problems that railroads have faced since the earliest days is the prevention of service failures in track. As is the case with all modes of high-speed travel, failures of an essential component can have serious consequences. The North American railroads have been inspecting their costliest infrastructure asset, the rail, since the late 1920s. With increased traffic at higher speed, and with heavier axle loads in the 1990s, rail inspection is more important today than it has ever been. Although the focus of the inspection seems like a fairly well-defined piece of steel, the testing variables present are significant and make the inspection process challenging.

Rail inspections were initially performed solely by visual means. Of course, visual inspections will only detect external defects and sometimes the subtle signs of large internal problems. The need for a better inspection method became a high priority because of a derailment at Manchester, NY in 1911, in which 29 people were killed and 60 were seriously injured. In the U.S. Bureau of Safety's (now the National Transportation Safety Board) investigation of the accident, a broken rail was determined to be the cause of the derailment. The bureau established that the rail failure was caused by a defect that was entirely internal and probably could not have been detected by visual means. The defect was called a transverse fissure (Fig. 6–60). The railroads began investigating the prevalence of this defect and found transverse fissures were widespread.

Fig. 6–60　Transverse Fissure

One of the methods used to inspect rail is ultrasonic inspection. Both normal- and angle-beam techniques are used, as are both pulse-echo and pitch-catch techniques. The different transducer arrangements offer different inspection capabilities. Manual contact testing is done to evaluate small sections of rail but the ultrasonic inspection has been automated to allow inspection of large amounts of rail.

Fluid filled wheels or sleds are often used to couple the transducers to the rail. Sperry Rail Services, which is one of the companies that perform rail inspection, uses Roller Search Units (RSU's) comprising a combination of different transducer angles to achieve the best inspection possible. A schematic of an RSU is shown in Fig. 6–61.

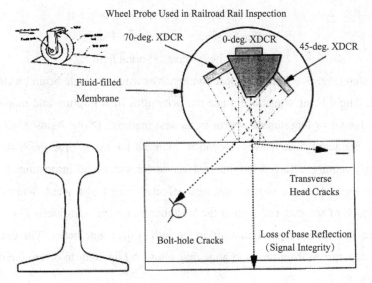

Fig. 6-61 Rail Inspections

■ 〖Point 2〗 Weldments (Welded Joints)

The most commonly occurring defects in welded joints are porosity, slag inclusions, lack of side-wall fusion, lack of inter-run fusion, lack of root penetration, undercutting and longitudinal or transverse cracks.

With the exception of single gas pores all the defects listed are usually well detectable by ultrasonics. Most applications are on low-alloy construction quality steels, however, welds in aluminum can also be tested. Ultrasonic flaw detection has long been the preferred method for non-destructive testing in welding applications. This safe, accurate and simple technique has pushed ultrasonics to the forefront of inspection technology.

Ultrasonic weld inspections are typically performed using a straight beam transducer in conjunction with an angle beam transducer and wedge. A straight beam transducer, producing a longitudinal wave at normal incidence into the test piece, is first used to locate any laminations in or near the heat-affected zone (Fig. 6-62). This is important because an angle beam transducer may not be able to provide a return signal from a laminar flaw.

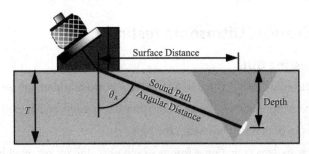

Fig. 6-62 Step 1

θ_R=Angle of Refraction

T=Material Thickness
Surface Distance=$\sin\theta_R$×Sound Path
Depth (1st Leg) =$\cos\theta_R$×Sound Path

The second step (Fig. 6–63) in the inspection involves using an angle beam transducer to inspect the actual weld. Angle beam transducers use the principles of refraction and mode conversion to produce refracted shear or longitudinal waves in the test material. (Note: Many AWS inspections are performed using refracted shear waves. However, material having a large grain structure, such as stainless steel may require refracted longitudinal waves for successful inspections.) This inspection may include the root, sidewall, crown, and heat-affected zones of a weld. The process involves scanning the surface of the material around the weldment with the transducer. This refracted sound wave will bounce off a reflector (discontinuity) in the path of the sound beam. With proper angle beam techniques, echoes returned from the weld zone may allow the operator to determine the location and type of discontinuity.

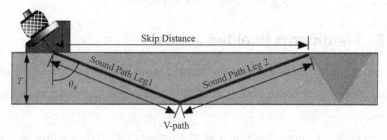

Fig. 6–63   Step 2

θ_R=Refracted Angle
T=Material Thickness
Skip=Distance=$2T \times \tan\theta_R$

$\text{Leg} = \dfrac{T}{\cos\theta_R}$

$\text{V-Path} = \dfrac{2T}{\cos\theta_R}$

To determine the proper scanning area for the weld, the inspector must first calculate the location of the sound beam in the test material. Using the refracted angle, beam index point and material thickness, the V-path and skip distance of the sound beam is found. Once they have been calculated, the inspector can identify the transducer locations on the surface of the material corresponding to the crown, sidewall and root of the weld.

〖Point 3〗 Advanced Ultrasonic Testing

Guided Wave Testing GUL

Guided wave ultrasonic testing (Fig. 6–64) detects corrosion damage and other defects over long (10–50 m) distances in piping. A special tool (transducer ring) is clamped around the pipe and transmits guided waves in both directions along the pipe.

Reflected signals from defects and pipe features such as welds are received by the transducer ring and sent to the main unit. Sophisticated processing and analysis software allows trained operators to interpret these signals and report their findings.

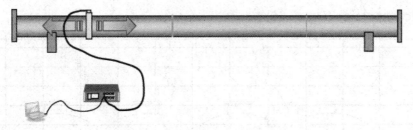

Fig. 6–64   Guided Wave Ultrasonic Testing

Developments to the technology since its introduction in the 1990s have lead to significant gains in the quality and reliability of test data and better tools for its interpretation. The resulting GUL G3 system operated by Matrix is the most advanced system of its type and offers:

•Tooling designed for easy attachment and use.

•Substantially automated testing: tool and pipe size information is read by the G3 unit to minimise manual operator input. The system checks and calibrates the probes before and after each test to ensure each data set is valid.

Ultrasonic signals (Fig. 6–65) swept over the full frequency range, in both directions, in a single shot. In addition to faster, more efficient data acquisition, having all the data in a single file makes analysis faster and more certain. Special, wide frequency probes are available for cases such as buried pipe where attenuation is too high over the standard frequency range.

Fig. 6–65   Aplication of Guided Wave Ultrasonic Testing

•Sophisticated software routines help identify and classify signals from the pipe.

•Embedded reporting software allows the operator to analyse the results and produce a report on the spot.

•Enhanced Focusing Capability (EFC) rings (pipes 4" and above) improve defect characterisation and provides colour coded C-scan type maps of the pipe. In effect, the EFC processing focuses on all reflectors over the entire range of the shot. This improves sensitivity as well as making the guided wave data more understandable to the end user.

•Rigorous operator training and certification with individual electronic keys that activate the system and track its use by each operator.

Coustic Inspection (Fig. 6-66)

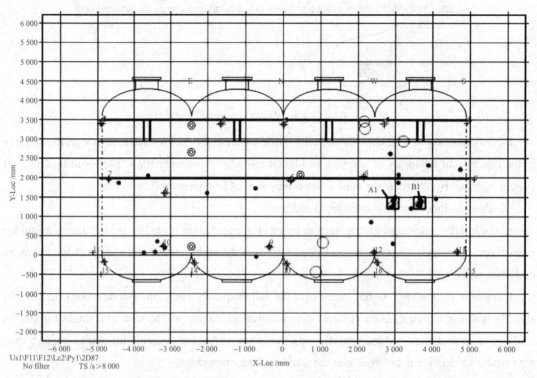

Fig. 6-66 Coustic Inspection

Acoustic Emission (AE) is high frequency sound generated by cracks and similar flaws in materials when stressed. Acoustic Emission Testing generally requires loading of a vessel or piping by filling or a pressure increase for detection of cracks and other defects.

For most in-service equipment, the requirement is to increase the pressure or level by 5% to 10% over the operating level while monitoring and recording AE activity. In some special applications such as on-line monitoring and storage tank floor assessment the AE signals are generated by the environmental conditions and no additional loading is required.

AE signals generated by the test are detected by surface mounted, high frequency acoustic sensors. The test procedure and AE data evaluation are designed to detect damage and provide a measure of a vessel's response to the applied load. Structurally significant defects produce relatively intense acoustic emission activity when subjected to additional stress from filling or pressure. There are many applications for AE testing on process plant (Fig. 6-67) including:

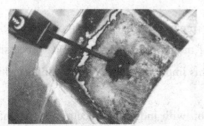

Fig. 6-67 Application of Coustic Inspection

AE testing of pressure vessels, tanks and piping is a means of detecting in-service damage such as stress corrosion cracking. The test requires that a vessel is pressurised or filled to five to ten percent above its operating level as it is monitored with AE equipment. AE activity is evaluated and graded in terms of its approximate location and intensity. Further testing or inspection of specific areas may be required if significant AE activity is detected.

Refrigerated Ammonia Tank testing with AE can avoid the costs and downtime of an internal inspection. AE testing is listed as an applicable non-intrusive test method in the European Fertiliser Manufacturers Association guidelines. This provides for use of AE where an internal inspection has been carried out previously. An external AE test is not only less costly than an internal inspection; it also avoids the risk of damaging the tank through oxygen contamination and thermal stress during recommissioning. These factors have led to increased use of AE testing for ammonia tanks.

AE Testing of Composite (FRP, GRP) Tanks, Piping and Vessels, is effective and relatively simple. In the case of a fibreglass tank, the test requires attaching AE sensors spaced 1.5 to 2 m (5 to 7 ft) apart and filling with the process fluid (or water) after a period at reduced level. Evaluation and grading criteria are applied and the results reported in terms of areas of significant AE activity and any follow-up actions required.

AE Tank floor testing is a means of assessing corrosion activity in aboveground storage tank floors. AE signals from on-going corrosion of the floor are transmitted through the fluid and detected by sensors placed around the tank shell. The test is carried out under static conditions and aims to grade the corrosion activity from "I" (little or no active damage) to "IV" (very active damage) . A grade I tank would normally be retested in 4-years or so whereas a grade IV result suggests the need for near term internal inspection. This test is best used to survey a number of tanks to help set inspection priorities and avoid unnecessary tank entries.

Remote on-line monitoring with AE is an option for cases where a discrete AE (pressure) test is not feasible or suitable for detecting process related damage. In such cases, AE sensors are installed and connected to instrumentation on site and controlled through an Internet connection. Thermocouple, pressure and other process data are recorded with the AE data. The monitoring process is controlled and evaluated remotely from a IRISNDT office.

Cooldown Testing uses thermal stress generated in piping and vessels as a unit is shut down. The test uses sensors mounted on acoustic waveguides to detect acoustic emission as the unit is shut down and cooled. Evaluation of the AE data helps focus inspection work on areas producing significant AE activity.

Although an AE test detects defects (Fig. 6–68), it does not determine their type, size or exact position. The main purpose and benefit of an AE test is to determine if there is a structural problem,

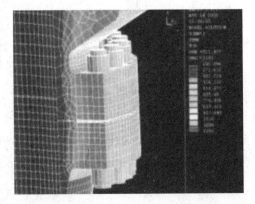

Fig. 6–68 AE Test Detects Defects

approximately where it is and give a measure of its severity. A complementary inspection method such as (shear wave) ultrasonics is needed to map out and size any flaws.

Limitations

Qualitative assessment, does not characterise a flaws or corrosion damage in precise (type of flaw, orientation, size, depth) terms.

Generally not sensitive to fabrication defects such as porosity, slag inclusions and small lack of fusion defects in operating equipment.

Background noise and other properties of a particular vessel may limit detection sensitivity.

Weld Inspection (Pulse-echo and TOFD)

The scanners and Microtomo system are capable of performing both conventional pulse-echo (angle beam) and Time of Flight Diffraction (TOFD) scans: simultaneously if required. In this case, the welds of a reactor were examined to detect and size flaws. The welds were examined using a 45 deg pulse-echo technique. Data was recorded in blocks (scans) anywhere from 12" to 210" in length. Successive scans began at the endpoint of the previous scan. To ensure complete coverage of the examination volume, a data point and depth reading was taken every 0.04" in the axial (Y) scan direction before indexing by 0.200" in the circumferential or longitudinal (X) direction to begin the next scan stroke. This (better than 50%) overlap of the search unit while indexing ensured that small flaws could be detected reliably. The longitudinal weld seam L2-2 was later examined by automated TOFD technique for better sizing of the indications. 60 deg longitudinal wave transducers were used for this examination. The Pulse-echodata is shown in Fig. 6-69.

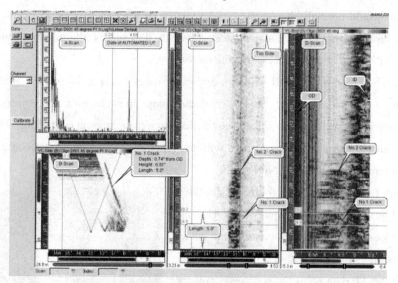

Fig. 6–69   Pulse-echo Data

TOFD indications (Fig. 6–70) are shown by a top view of stacked A-scans. The grey shade indicates amplitude, vertical position corresponds to position along the weld and the horizontal axis shows time (of flight) . These elements allow location and sizing of defects.

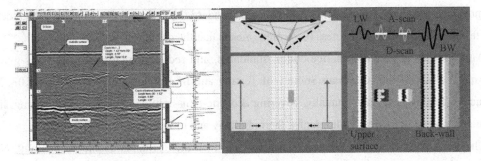

Fig. 6–70   TOFD Indications

Phased Array (PA)

A single phased array probe can simulate many different conventional probes. The ultrasonic beam may be both steered (change of angle), focussed and scanned without moving the probe. Sound beams of many angles can be generated sequentially, inspecting a large portion of the component's cross-section.

To achieve this same coverage using conventional techniques, multiple probes would have to be moved over the component's surface. Advantages of the phased array are that it allows inspection of complex geometries with a single probe and scan times are greatly reduced. Fig. 6–71 shows a 45 to 70-degree scan of a butt weld. The dark gray areas on the C-scan view are from lack of fusion defects.

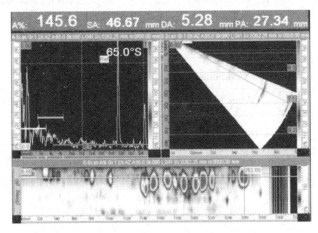

Fig. 6–71   Phased Array (PA)

IRISNDT have both the RD-Tech Omniscan and the Focus Phased Array Systems.

Put into Practice

1. The following essay explains the principle of defect detection.

What are standards?

Standards are documented agreements containing technical specifications or other precise criteria to be used consistently as rules, guidelines or definitions of characteristics, in order to ensure that materials, products, processes and services are fit for their purpose.

For example, the format of the credit cards, phone cards, and "smart" cards that have become commonplace is derived from an ISO International Standard. Adhering to the standard, which defines such features as an optimal thickness (0.76 mm), means that the cards can be used worldwide.

An important source of practice codes, standards, and recommendations for NDT is given in the Annual Book of the American Society of Testing and Materials, ASTM, Volume 03.03. Non-destructive Testing is revised annually, covering acoustic emission, eddy current, leak testing, liquid penetrants, magnetic particle, radiography, thermography and ultrasonics.

There are many efforts on the part of the National Institute of Standards and Technology (NIST) and other standards organizations, both national and international, to work through technical issues and harmonize national and international standards.

2. Translate the inspection reports in Table 6-2.

Table 6-2 Material UT Report

SH/T 3503-J124-1		Material UT Report (I) Page of		Project: Unit:	
Entrusted by		Report No.		Certificate No.	
Contractor		Test Piece Description		Test Piece Size	mm
Test Piece Material		Test Criteria		Test Percentage	%
Qualified Level		Test Piece Type		HT Status	
Test Surface		Surface Condition		Standard Test Coupon	
Equipment Model		Probe Model		Surface Compensation	dB
Scanning Percentage		Test Sensitivity		Couplant	
Test Position No.	Defect No.	Defect Area /cm²	Equivalent Diameter /mm	Buried Depth /mm	Evaluated Level	Remarks
Tested by: Qualification: UT Level		Reviewed by: Qualification: UT Level		Inspection agency: (Seal) Report Date:	

Words and Phrases

ultrasonic [ˌʌltrəˈsɒnɪk] *adj.* 超声的 *n.* 超声波   surface wave 表面波

Rayleigh Wave 瑞利（表面）波
guided wave 导波
Lamb wave 兰姆波
ultrasonic wave 超声波
Curie 居里［姓氏］放射性强度单位 $=3.7 \times 10^{10}$（Bq）衰
piezoelectric adj.［物］压电的
piezoelectric effect 压电效应
signal-to-noise ratio (STNR) 信噪比
transducer [trænz'djuːsə] n. 换能器，传感器，变换器
transducer arrays 换能器阵列
pulse-echo 脉冲回波
supersonic [ˌsuːpə'sɒnɪk] adj. 超声波的 n. 超声波，超声频
reflectoscope n.［医］反射灯，投射灯，超声波探伤仪（反射测试仪）
immersion n. 沉浸
microscope ['maɪkrəˌskəʊp] n. 显微镜
microprocessor [ˌmaɪkrəʊ'prəʊsesə] n.［计］微处理器
equilibrium n. 平衡，平静，均衡
oscillate ['ɒsɪˌleɪt] v. 振荡
transverse wave 横波
longitudinal wave 纵波
crest [krest] n. 波峰
valley ['væli] n. 波谷
wave front 波前，波阵面
shear wave 切变波
shear stress 切应力
phase velocity 相速度
viscosity [vɪ'skɒsəti] n. 黏质，黏性
pressure wave 压力波
rarefaction [ˌreərɪ'fæʃn] n. 变稀薄，稀薄
sinusoid ['saɪnəˌsɔɪd] n. 正弦波，正弦曲线
harmonic (sinusoidal) waves 谐波（正弦波）
scalar ['skeɪlə] n. 数量，标量
vector quantity 矢量

dot product 点积，点乘，标量积，数量积
gain [geɪn] n. 增益（系数），放大（系数，率）
attenuate [ə'tenjueɪt] v. 衰减
major axis （椭圆的）长轴
minor axis （椭圆的）短轴
symmetric [sɪ'metrɪk] adj. 相称性的，均衡的
antisymmetric adj. 反对称的
impedance n. 阻抗
acoustic impedance 声阻抗
critical angle n.［物］临界角
real number n.［数］实数
complex number ［数］复数
imaginary number 虚数
plate wave 板波
resonant wave 共振波
perspex n. 胶质玻璃，有机玻璃
polyethylene [ˌpɒlɪ'eθəliːn] n. 聚乙烯
plexiglass ['pleksɪˌglɑːs] n. 树脂玻璃
plexiglass wedge 有机玻璃斜楔
nondispersive [ˌnɒndɪs'pɜːsiv] 非分散的
isotropic [ˌaɪsəʊ'trɒpɪk] adj. 全（无）向性的，各向同性的
decibel ['desɪbel] n. 分贝
neper n.［物］奈培（衰减单位，等于 8.686 分贝）
side lobe 旁瓣，副瓣
near field or near zone or Fresnel zone 近场
far field or far zone, Fraunhofer zone 远场
continuous wave (CW) 连续波
diffract [dɪ'frækt] vt. 使分散，衍射
oscilloscope [ə'sɪləskəʊp] n.［物］示波器
millisecond ['mɪliˌsekənd] n. 毫秒
pulser ['pʌlsə] n. 脉冲发生器
synchronize v. 同步
astigmatism [ə'stɪgmətɪzəm] n. 散光
curvature ['kɜːvətʃə] n. 弯曲，曲率
normal incidence 正入射，法线入射，垂直入射
rise time 脉冲前沿，上升时间

single crystal 单晶（体）
quartz [kwɔːts] n. 石英
ferroelectric [ˌferəʊɪˈlektrɪk] n. ［电］铁电物质 adj. 铁电的
polycrystalline [ˌpɒlɪˈkrɪstəˌlaɪn] adj. ［物］多晶的
Young's module 杨氏模量
Poisson's ratio 泊松比
modulus [ˈmɒdjʊləs] n. 系数，模数
dual crystal transducer 双晶探头
damping [ˈdæmpɪŋ] n. 阻尼，减幅，衰减
die down v. 变弱，逐渐停止，渐渐消失
dead zone 盲区

backing material 背衬材料
wear plate （探头的）保护膜，防磨板
alumina n. ［化］氧化铝（矾土）
titanium oxide 氧化钛
epoxy resin n. 环氧树脂
couplant n. 耦合剂
timer-trigger circuit 时钟触发器电路
reference test object 参考试块
time base 时基（线）
two-dimensional 二维的
normal beam transducer 直探头
angle beam transducer 斜探头
angle beam incidence 斜入射

 A Sheet Work Manual

Ultrasonic Testing Procedure

1. Procedure

(1) Testing time.

① Supersonic testing should be arranged at the temperature of weld cooling to that of surroundings.

② The weld with crack extension should be done after 48 hours of welding.

(2) Base material and groove.

① The groove in the working drawing should be confirmed together with welding method at site, such as angle of groove, which is based on the confirmation of gap and plate thickness before testing.

② Weld reinforcement of cover and geometrical of penetration bead should be checked prior to the inspection to avoid the misjudgment.

(3) Preparation of testing surface.

Testing sensitivity of angular probe should not be lower than assessment sensitivity. When detecting horizontal defect, the testing sensitivity should be promoted by 6 dB; when straight probe tests weld, the testing sensitivity should be Φ_2 flat bottom hole.

(4) Scavenge.

① The scavenge technique can be applied to all the weld beads and make the comparison between one side of weld seam with at least 10 mm and hot affecting area, and which can cover the bigger area.

② The suitable scavenge should be applied to all the testing weld seam to make sure that the location, direction, shape of the defect as well as make clear between defective signal and fake signal. The normal scavenge can be seen in Fig. 6-72.

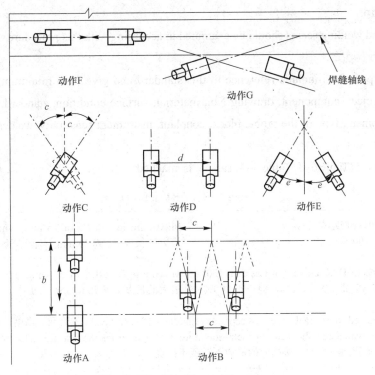

Fig. 6-72 The Normal Scavenge

③ The speed of scavenge should not exceed than 150 mm/s, the interval between the movement of two probes should be at least the 10% overlapping of the width of probe.

④ Reflection echo with more than 20% DAC defective amplitude should be judged whether it is defective according to the moving position of probe, direction and the location of reflection echo.

⑤ Scavenge of vertical defects. The direction of probe should be perpendicular to the axis of weld seam and to do the wave scavenge along the testing route.

⑥ Scavenge of horizontal defects. Probe should do the parallel scavenge along the weld axis on the weld seam, or do the inclined parallel scavenge, or do the inclined parallel scavenge on angle less than 15° between the sides of weld seam and axis.

(5) Measurement of defect location, indicating length.

① Measurement of defect location. When found the defect, measurement of defect location should be subject to the location of the biggest reflection echo with found defects.

② Measurement of indicating length. When the defective reflection echo only has one peak, the 6 dB method should be applied to measure the length.

When the defective reflection echo has many peaks, the endpoint 6 dB method should be applied to measure the length.

2. Evaluation levels

Echo height 回波高度	Acceptance criteria for ultrasonic testing see Note 超声波检测的合格标准（见备注）

3. Repairing

The repaired welding seams should be examined in according with this procedure.

4. Testing report

The test report shall include reference to this standard and give, as a minimum, the following information: project, component, drawing No., material, surface condition, standard, test date, type of testing equipment, type of the probe, block, couplant, instrument sensitivity, welder, sketch of test position, defect.

The test report (Table 6-3, Table 6-4) sample is attached:

Table 6-3　Test Report1

Greater than 100% of DAC curve DAC 曲线大于 100%	Maximum length $t/2$ or 25 mm, whichever is less 最大的长度 $t/2$ 或 25 mm, 或者更短
Greater than 50% of DAC curve, but less than 100% of DAC curve DAC 曲线大于 50%，但小于 100%	Maximum length t or 50 mm, whichever is less 最大的长度 t 或 50 mm, 或者更小
Indications evaluated to be cracks are unacceptable regardless of echo height 不管回波高度，显示的裂缝不接受 Indications evaluated to be lack of penetration in joints welded from one side are unacceptable regardless of echo height. 不管回波高度，接缝处的贯通不够就不接受	
NOTE 备注： Two adjacent individual discontinuities of length L_1 and L_2 situated on a line and where the distance L between them is shorter than the shortest discontinuity are to be regarded as a continuous discontinuity of length L_1+L+L_2，位于一条线上的 L_1 和 L_2 长度的两个相邻的不连续性，以及它们之间的距离比最短的不连续性要短，这被认为是 L_1+L+L_2 长度的无间断的不连续性	

项目 (Project)：	工程编号 (Hull No.)：	探伤日期 (Test date)：
评定标准 (Identify Standard)：	合格等级 (Accept Level)：	探测表面状态 (Test Surface)：
耦合剂 (Couplant)：	仪器型号 (Detector Type)：	探头频率 (Probe Frequency)：
折射角 (Refraction Angle)：	探伤灵敏度 (Instrument Sensitivity)：	补偿 (Transfer)：
灵敏度试块 (Sensitivity of Block)：	材质 (Material)	检测灵敏度 (Inspection Sensitivity)
厚度 (Thickness)：		施工单位 (Work's Shop)：
探伤部位 (Position of Detection)：		
图纸编号 (Drawing No.)：		
结论 (Conclusion)：		

Table 6-4 Test Report 2

序号 No.	结构名称 Structure Member	焊缝编号 Welds No.	板厚 Thickness /mm	缺陷 Discontinuity					评定级别 Assessing Grade	备注 Remark
				波辐 Amplitude /dB	指示长度 Indication Length /mm	坐标位置 Coordinates		深度 Depth /mm		
						X/mm	Y/mm			
检测人员 Inspector			审核人员 Authorized by			验船师 Surveyor				

Item 07 Eddy Current Testing

Learning Objectives

1. Knowledge objectives
(1) To grasp the words, related terms and abbreviations about ET.

(2) To grasp the classification about ET system.

(3) To know the Instruments and Equipment of ET system.

(4) To know the testing procedure of ET system.

2. Competence objectives
(1) To be able to read and understand frequently used & complex sentence patterns, capitalized English materials and obtain key information quickly.

(2) To be able to communicate with English speakers about the topic freely.

(3) To be able to fill in the job cards in English.

3. Quality objectives
(1) To be able to self-study with the help of aviation dictionaries, the Internet and other resources.

(2) To do a good job of detection of safety protection.

7.1 Introduction of Eddy Current Testing

〖Point 1〗 Eddy Current Testing

Eddy current testing (Fig. 7-1) is used to detect surface and near-surface flaws in conductive materials.

This non-destructive testing technique is non-hazardous and commonly used in industries such as aerospace, rail, automotive, marine and manufacturing. One of the major advantages of eddy current testing is that inspection requires minimum preparation as there is no need to remove surface paint or coating. This makes it suitable for inspecting painted structures, parts and components.

Fig. 7-1 Eddy Current Testing

Eddy current testing equipment is highly portable, reliable and can detect very small cracks. Results are instant, ideal for on-site testing on-site and plant inspections. Flaws can be reported immediately to site and operation managers, allowing for quicker decision making. In addition, the portability of equipment means that we can inspect equipment or assets that are difficult to access and test complex shapes and sizes.

Eddy current testing uses electromagnetic induction to detect defects in both ferrous and non-ferrous materials by inducing an eddy current field in the specimen under test. A variety of inspections can be performed with eddy current testing and it is typically used for surface and near-surface flaw detection, metal and coating thickness measurement, and metal sorting by grade and hardness.

A specially designed coil energised with an alternating-current is placed in proximity to the test surface, generating a changing magnetic field that interacts with the test-part and produces eddy currents in the vicinity.

Variations in the changing phases and magnitude of these eddy currents are then monitored through the use of a receiver-coil or by measuring changes to the alternate current flowing in the primary excitation-coil.

The electrical conductivity variations, the magnetic permeability of the test-part or the presence of any discontinuities, will cause a change in the eddy current and a corresponding change in phases and amplitude of the measured current. The changes are shown on a screen and are interpreted to identify defects.

▮ 〖Point 2〗 Work Process

The process relies upon a material characteristic known as electromagnetic induction. When an alternating current is passed through a conductor, a copper coil for example, an alternating magnetic field is developed around the coil and the field expands and contracts as the alternating current rises and falls. If the coil is then brought close to another electrical conductor, the fluctuating magnetic field surrounding the coil permeates the material and, by Lenz's Law, induces an eddy current to flow in the conductor. This eddy current, in turn, develops its own magnetic field. This "secondary" magnetic field opposes the "primary" magnetic field and thus affects the current and voltage flowing in the coil.

Any changes in the conductivity of the material being examined, such as near-surface defects or differences in thickness, will affect the magnitude of the eddy current. This change is detected using either the primary coil or the secondary detector coil, forming the basis of the eddy current testing inspection technique.

Permeability is the ease in which a material can be magnetised. The greater the permeability the smaller the depth of penetration. Non-magnetic metals such as austenitic stainless steels, aluminium and copper have very low permeability, whereas ferritic steels have a magnetic permeability several hundred times greater.

Eddy current density is higher, and defect sensitivity is greatest, at the surface and this decreases with depth. The rate of the decrease depends on the "conductivity" and "permeability" (Fig. 7-2) of the metal. The conductivity of the material affects the depth of penetration. There is a greater flow of eddy current at the surface in high conductivity metals and a decrease in penetration in metals such as copper and aluminium.

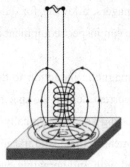

Fig. 7-2 The "Conductivity" and "Permeability"

The depth of penetration may be varied by changing the frequency of the alternation current—the lower the frequency, the greater depth of penetration. Therefore, high frequencies can be used to detect near-surface defects and low-frequencies to detect deeper defects. Unfortunately, as the frequency is decreased to give greater penetration, the defect detection sensitivity is also reduced. There is therefore, for each test, an optimum frequency to give the required depth of penetration and sensitivity.

Signal from a defect-free sample in Fig. 7-3.

Fig. 7-3 Signal from a Defect-free Sample

■ 〖Point 3〗 History of Eddy Current Testing

Eddy current testing has its origins with Michael Faraday's discovery of electromagnetic induction in 1831. Faraday (Fig. 7-4) was a chemist in England during the early 1800s and is credited with the discovery of electromagnetic induction, electromagnetic rotations, the magneto-optical effect, diamagnetism and other phenomena. In 1879, another scientist named Hughes recorded changes in the properties of a coil when placed in contact with metals of different conductivity and permeability. However, it was not until the Second World War that these effects were put to practical use for testing materials. Much work was done in the 1950s and 1960s, particularly in the aircraft and nuclear industries. Eddy current testing is now a widely used and well-understood inspection technique.

Fig. 7-4 Michael Faraday (1791—1867)

■ 〚Point 4〛 Advantages and Limitations

Advantages
- Able to detect surface and near-surface cracks as small as 0.5 mm.
- Able to detect defects through several layers, including non-conductive surface coatings, without interference from planar defects.
- Non-contact method making it possible to inspect high-temperature surfaces and underwater surfaces.
- Effective on test objects with physically complex geometries.
- Provides immediate feedback.
- Portable and light equipment.
- Quick preparation time—surfaces require little pre-cleaning and couplant is not required.
- Able to the measure electrical conductivity of test objects.
- Can be automated for inspecting uniform parts such as wheels, boiler tubes, or aero-engine disks.

Limitations
- Can only be used on conductive materials.
- The depth of penetration is variable.
- Very susceptible to magnetic permeability changes—making testing of welds in ferromagnetic materials difficult—but with modern digital flaw detectors and probe design, not impossible.
- Unable to detect defects that are parallel to the test object's surface.
- Careful signal interpretation is required to differentiate between relevant and non-relevant indications.

■ 〚Point 5〛 Applications (Fig. 7–5)

- Inspection of parts or components including:

    Welded joints

    Bores of in-service tubes

    Bores of bolt holes

    Metal tubes

    Friction stir welds

    Gas turbine blades

    Nozzle welds in nuclear reactors

    Hurricane propeller hubs

    Cast iron bridges

    Gas turbine blades

- Detection of defects including:

    Surface-breaking defects

    Linear defects (as small as 0.5 mm deep and 5 mm long)

Cracks

Lack of fusion

Generalised corrosion (particularly in the aircraft industry for the examination of aircraft skins)

• Other applications:

Identification of both ferrous and non-ferrous metals and with certain alloys—in particular the aluminium alloys

Establishing the heat treatment condition

Determining whether a coa ting is non-conductive

Heat treat verification of metals

External Thread Crack Detection Using Eddy Current   Eddy Current Testing of Pipelines

Fig. 7–5　Applications

Put into Practice

1. Fig. 7–6 explains the principle of defect detection.

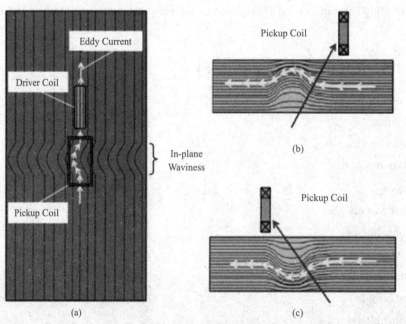

Fig. 7–6　The Principle of Defect Detection

2. Read Fig. 7–7 to illustrate the relationship between the frequency of detection and the depth of detection of defects.

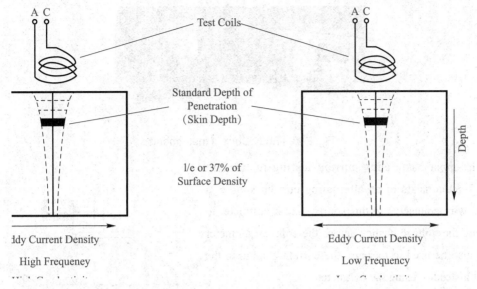

Fig. 7-7 Relationship between the Frequency and the Depth

3. The following essays summarize the advantages of vortex detection.

One of the major advantages of eddy current as an NDT tool is the variety of inspections and measurements that can be performed. In the proper circumstances, eddy currents can be used for: crack detection; material thickness measurements; coating thickness measurements; conductivity measurements; material identification; heat damage detection; case depth determination; heat treatment monitoring.

Some of the advantages of eddy current inspection include: sensitive to small cracks and other defects, detects surface and near surface defects; inspection gives immediate results; equipment is very portable; method can be used for much more than flaw detection; minimal part preparation is required; test probe does not need to contact the part; inspects complex shapes and sizes of conductive materials.

7.2 Instrumentation

〖Point 1〗 Eddy Current Instruments

Eddy current instruments (Fig. 7-8) can be purchased in a large variety of configurations. Both analog and digital instruments are available. Instruments are commonly classified by the type of display used to present the data. The common display types are analog meter, digital readout, impedance plane and time versus signal amplitude. Some instruments are capable of presenting data in several display formats.

Fig. 7-8 Eddy Current Instruments

The most basic eddy current testing instrument (Fig. 7-9) consists of an alternating current source, a coil of wire connected to this source, and a voltmeter to measure the voltage change across the coil. An ammeter could also be used to measure the current change in the circuit instead of using the voltmeter.

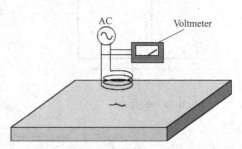

Fig. 7-9 Basic Eddy Current Testing Instrument

While it might actually be possible to detect some types of defects with this type of equipment, most eddy current instruments are a bit more sophisticated. In the following pages, a few of the more important aspects of eddy current instrumentation will be discussed.

〖Point 2〗 Resonant Circuits

Eddy current probes typically have a frequency or a range of frequencies that they are designed to operated. When the probe is operated outside of this range, problems with the data can occur. When a probe is operated at too high of a frequency, resonance can occurs in the circuit. In a parallel circuit with resistance (R), inductance (X_L) and capacitance (X_C), as the frequency increases X_L decreases and X_C increase. Resonance occurs when X_L and X_C are equal but opposite in strength. At the resonant frequency, the total impedance of the circuit appears to come only from resistance since X_L and X_C cancel out. Every circuit containing capacitance and inductance has a resonant frequency that is inversely proportional to the square root of the product of the capacitance and inductance.

In eddy current probes and cables, it is commonly stated that capacitance is negligible. However, even circuits not containing discreet components for resistance, capacitance, and inductance can still exhibit their effects. When two conductors are placed side by side, there is always some capacitance between them. Thus, when many turns of wire are placed close together in a coil, a certain amount of stray capacitance is produced. Additionally, the cable used to interconnect pieces of electronic equipment or equipment to probes, often has some capacitance, as well as, inductance. This stray capacitance is usually very small and in most cases has no significant effect. However, they are not negligible in sensitive circuits and at high frequencies they become quite important.

The applet below represents an eddy current probe with a default resonant frequency of about 1.0 kHz. An ideal probe might contain just the inductance, but a realistic probe has some resistance and some capacitance. The applet initially shows a single cycle of the 1.0 kHz current passing through the inductor.

〖Point 3〗 Bridges

The bridge circuit shown in the applet below is known as the Maxwell-Wien bridge (often called the Maxwell bridge), and is used to measure unknown inductances in terms of calibrated resistance and capacitance. Calibration-grade inductors are more difficult to manufacture than capacitors of similar precision, and so the use of a simple "symmetrical" inductance bridge is not always practical. Because the phase shifts of inductors and capacitors are exactly opposite each other, a capacitive impedance can balance out an inductive impedance if they are located in opposite legs of a bridge, as they are here.

Unlike this straight Wien bridge, the balance of the Maxwell-Wien bridge is independent of the source frequency. In some cases, this bridge can be made to balance in the presence of mixed frequencies from the AC voltage source, the limiting factor being the inductor's stability over a wide frequency range.

Exercise:

Using the equations within the applet, calculate appropriate values for C and R_2 for a set of probe values. Then, using your calculated values, balance the bridge. The oscilloscope trace representing current (brightest green) across the top and bottom of the bridge should be minimized (straight line).

In the simplest implementation, the standard capacitor (C) and the resistor in parallel with it are made variable, and both must be adjusted to achieve balance. However, the bridge can be made to work if the capacitor is fixed (non-variable) and more than one resistor is made variable (at least the resistor in parallel with the capacitor, and one of the other two). However, in the latter configuration it takes more trial-and-error adjustment to achieve balance as the different variable resistors interact in balancing magnitude and phase.

Another advantage of using a Maxwell bridge to measure inductance rather than a symmetrical inductance bridge is the elimination of measurement error due to the mutual inductance between the two inductors. Magnetic fields can be difficult to shield, and even a small amount of coupling between coils in a bridge can introduce substantial errors in certain conditions. With no second inductor to react within the Maxwell bridge, this problem is eliminated.

〖Point 4〗 Complex Impedance Plane (Fig. 7–10): Eddy Current Scope

Phase Angle: The difference in phase between two sinusoidally varying quantities.

Capacitive Reactance: a property of a circuit containing capacitance that together with any resistance makes up its impedance.

Inductive Reactance: a property of a circuit containing inductance that together with any resistance makes up its impedance.

Electrical Impedance (Z) is the total opposition that a circuit presents to an alternating current. Impedance, measured in ohms, may include resistance (R), inductive reactance (X_L) and capacitive reactance (X_C). Eddy current circuits usually have only R and (X_L) components. As discussed in the page on impedance, the resistance component and the reactance component are not in phase, so vector addition must be used to relate them with impedance. For an eddy current circuit with resistance and inductive reactance components, the total impedance is calculated using the following equation.

You will recall that this can be graphically displayed using the impedance plane diagram as seen above. Impedance also has an associated angle, called the phase angle of the circuit, which can be calculated by the following equation.

$$Z=R+j(X_L-X_C)$$

The impedance plane diagram is a very useful way of displaying eddy current data. As shown in Fig. 7–11, the strength of the eddy currents and the magnetic permeability of the test material cause the eddy current signal on the impedance plane to react in a variety of different ways.

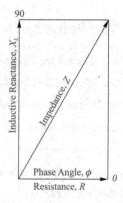

Fig. 7–10 Complex Impedance Plane

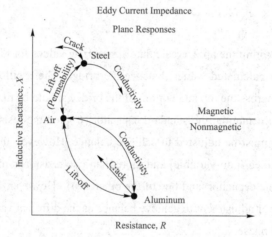

Fig. 7–11 The Impedance Plane Diagram

If the eddy current circuit is balanced in air and then placed on a piece of aluminum, the resistance component will increase (eddy currents are being generated in the aluminum and this takes energy away from the coil, which shows up as resistance) and the inductive reactance of the coil decreases (the magnetic field created by the eddy currents opposes the coil's magnetic field and the net effect is a weaker magnetic field to produce inductance) . If a crack is present in the material, fewer eddy currents will be able to form and the resistance will go back down and the inductive reactance will go back up. Changes in conductivity will cause the eddy current signal to change in a different way.

When a probe is placed on a magnetic material such as steel, something different happens. Just like with aluminum (conductive but not magnetic), eddy currents form, taking energy away from the coil, which shows up as an increase in the coils resistance. And, just like with the aluminum, the eddy currents generate their own magnetic field that opposes the coils magnetic field. However, you will note for the diagram that the reactance increases. This is because the magnetic permeability of the steel concentrates the coil's magnetic field. This increase in the magnetic field strength completely overshadows the magnetic field of the eddy currents. The presence of a crack or a change in the conductivity will produce a change in the eddy current signal similar to that seen with aluminum.

In the applet below, liftoff curves can be generated for several nonconductive materials with various electrical conductivities. With the probe held away from the metal surface, zero and clear the graph. Then slowly move the probe to the surface of the material. Lift the probe back up, select a different material and touch it back to the sample surface.

■ 〚Point 5〛 Analog Meter

Analog instruments are the simplest of the instruments available for eddy current inspections. They are used for crack detection, corrosion inspection or conductivity testing. These types of instruments contain a simple bridge circuit, which compares a balancing load to that measured on the test specimen. If any changes in the test specimen occur which deviate from normal you will see a movement on the instruments meter.

Analog meters such as the D'Arsonval design pictured in Fig. 7–12, must "rectify" the AC into DC. This is most easily accomplished through the use of devices called diodes. Without going into elaborate detail over how and why diodes work as they do, remember that they each act like a one-way valve for electrons to flow. They act as a conductor for one polarity and an insulator for another. Arranged in a bridge, four diodes will serve to steer AC through the meter movement in a constant direction.

Fig. 7–12   Analog Meter

An analog meter can easily measure just a few microamperes of current and is well suited for use in balancing bridges.

Put into Practice

1. Fig. 7–13 shows which components of the inspection system are.

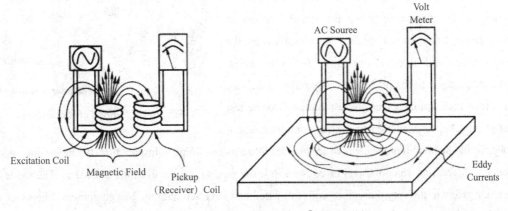

Fig. 7–13   The Inspection System

273

2. Translate the essays and read the diagram (Fig. 7-14) to summarize the principles of eddy current detection.

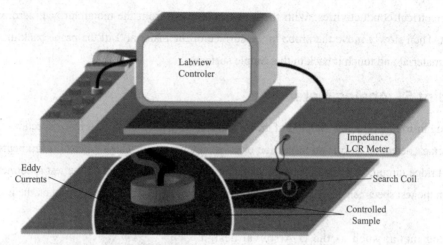

Fig. 7-14 The Eddy Current Testing (ECT) Method

The Eddy Current Testing (ECT) method is based on the interaction between a magnetic field coming from a search coil (the ECT sensor) and a tested material. This interaction induces eddy currents in the test piece. The distribution of these eddy currents is connected to the impedance of the search coil. By monitoring this impedance while scanning the controlled surface, defects can be geolocalized thanks to local impedance variations.

7.3 Probe/Coil Design

〖Point 1〗 Mode of Operation

Eddy current probes (Fig. 7-15) are available in a large variety of shapes and sizes. In fact, one of the major advantages of eddy current inspection is that probes can be custom designed for a wide variety of applications. Eddy current probes are classified by the configuration and mode of operation of the test coils. The configuration of the probe generally refers to the way the coil or coils are packaged to best "couple" to the test area of interest. An example of different configurations of probes would be bobbin probes, which are inserted into a piece of pipe to inspect from the inside out, versus encircling probes, in which the coil or coils encircle the pipe to inspect from the outside in. The mode of operation refers to the way the coil or coils are wired and interface with the test equipment. The mode of operation of a probe generally falls into one of four categories: absolute, differential, reflection and hybrid.

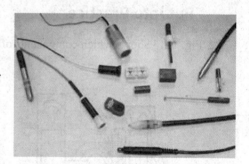

Fig. 7-15 Eddy Current Probes

Each of these classifications will be discussed in more detail below.

Absolute probes (Fig. 7–16) generally have a single test coil that is used to generate the eddy currents and sense changes in the eddy current field. As discussed in the physics section, AC is passed through the coil and this sets up an expanding and collapsing magnetic field in and around the coil. When the probe is positioned next to a conductive material, the changing magnetic field generates eddy currents within the material. The generation of the eddy currents take energy from the coil and this appears as an increase in the electrical resistance of the coil. The eddy currents generate their own magnetic field that opposes the magnetic field of the coil and this changes the inductive reactance of the coil. By measuring the absolute change in impedance of the test coil, much information can be gained about the test material.

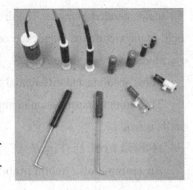

Fig. 7–16 Absolute Probes

Absolute coils can be used for flaw detection, conductivity measurements, liftoff measurements and thickness measurements. They are widely used due to their versatility. Since absolute probes are sensitive to things such as conductivity, permeability liftoff and temperature, steps must be taken to minimize these variables when they are not important to the inspection being performed. It is very common for commercially available absolute probes to have a fixed "air loaded" reference coil that compensates for ambient temperature variations.

Differential Probes

Move the probe to see the inspection results.

Differential probes have two active coils usually wound in opposition, although they could be wound in addition with similar results. When the two coils are over a flaw-free area of test sample, there is no differential signal developed between the coils since they are both inspecting identical material. However, when one coil is over a defect and the other is over good material, a differential signal is produced. They have the advantage of being very sensitive to defects yet relatively insensitive to slowly varying properties such as gradual dimensional or temperature variations. Probe wobble signals are also reduced with this probe type. There are also disadvantages to using differential probes. Most notably, the signals may be difficult to interpret. For example, if a flaw is longer than the spacing between the two coils, only the leading and trailing edges will be detected due to signal cancellation when both coils sense the flaw equally.

Reflection Probes

Reflection probes (Fig. 7–17) have two coils similar to a differential probe, but one coil is used to excite the eddy currents and the other is used to sense changes in the test material. Probes of this arrangement are often referred to as driver/pickup probes. The advantage of reflection probes is that the driver and pickup coils can be separately optimized

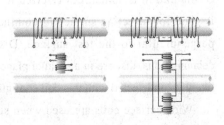

Fig. 7–17 Reflection Probes

for their intended purpose. The driver coil can be made so as to produce a strong and uniform flux field in the vicinity of the pickup coil, while the pickup coil can be made very small so that it will be sensitive to very small defects.

Some absolute and differential "transformer" type eddy current probes.

The through-transmission method is sometimes used when complete penetration of plates and tube walls is required.

Hybrid Probes (Fig. 7-18)

An example of a hybrid probe is the split D, differential probe shown to the right. This probe has a driver coil that surrounds two D shaped sensing coils. It operates in the reflection mode but additionally, its sensing coils operate in the differential mode. This type of probe is very sensitive to surface cracks. Another example of a hybrid probe is one that uses a conventional coil to generate eddy currents in the material but then uses a different type of sensor to detect changes on the surface and within the test material. An example of a hybrid probe is one that uses a Hall effect sensor to detect changes in the magnetic flux leaking from the test surface. Hybrid probes are usually specially designed for a specific inspection application.

Fig. 7-18 Hybrid Probes

■ 〖Point 2〗 Configurations

As mentioned at the begining of this section, eddy current probes are classified by the configuration and mode of operation of the test coils. The configuration of the probe generally refers to the way the coil or coils are packaged to best "couple" to the test area of interest. Some of the common classifications of probes based on their configuration include surface probes, bolt hole probes, inside diameter (ID) probes, and outside diameter (OD) probes.

Surface probes (Fig. 7-19) are usually designed to be handheld and are intended to be used in contact with the test surface. Surface probes generally consist of a coil of very fine wire encased in a protective housing. The size of the coil and shape of the housing are determined by the intended use of the probe. Most of the coils are wound so that the axis of the coil is perpendicular to the test surface. This coil configuration is sometimes referred to as a pancake coil and is good for detecting surface discontinuities that are oriented perpendicular to the test surface. Discontinuities, such as delaminations, that are in a parallel plane to the test surface will likely go undetected with this coil configuration.

Wide surface coils are used when scanning large areas for relatively large defects. They sample a relatively large area and

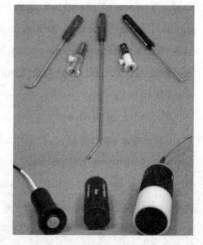

Fig. 7-19 Surface Probes

allow for deeper penetration. Since they do sample a large area, they are often used for conductivity tests to get more of a bulk material measurement. However, their large sampling area limits their ability to detect small discontinuities.

Pencil probes have a small surface coil that is encased in a long slender housing to permit inspection in restricted spaces. They are available with a straight shaft or with a bent shaft, which facilitates easier handling and use in applications such as the inspection of small diameter bores. Pencil probes are prone to wobble due to their small base and sleeves are sometimes used to provide a wider base.

Bolt Hole Probes

Bolt hole probes are a special type of surface probe that is designed to be used with a bolt hole scanner. They have a surface coil that is mounted inside a housing that matches the diameter of the hole being inspected. The probe is inserted in the hole and the scanner rotates the probe within the hole.

ID probes, which are also referred to as Bobbin probes or feed-through probes (Fig. 7-20), are inserted into hollow products, such as pipes, to inspect from the inside out. The ID probes have a housing that keep the probe centered in the product and the coil (s) orientation somewhat constant relative to the test surface. The coils are most commonly wound around the circumference of the probe so that the probe inspects an area around the entire circumference of the test object at one time.

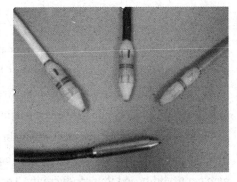

Fig. 7-20 ID/Bobbin Probes

OD probes are often called encircling coils (Fig. 7-21). They are similar to ID probes except that the coil (s) encircle the material to inspect from the outside in. OD probes are commonly used to inspect solid products, such as bars.

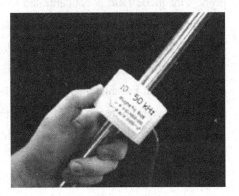

Fig. 7-21 OD Probes or Encircling Coils

〖Point 3〗 Shielding and Loading

One of the challenges of performing an eddy current inspection is getting sufficient eddy current field strength in the region of interest within the material. Another challenge is keeping the field away from non-relevant features of the test component. The impedance change caused by non-relevant features can complicate the interpretation of the signal. Probe shielding (Fig. 7-22) and loading are sometimes used to limit the spread and concentrate the magnetic field of the coil. Of course, if the magnetic field is concentrated near the coil, the eddy currents will also be concentrated in this area.

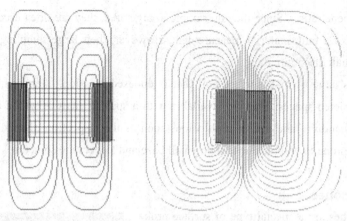

Fig. 7-22 Probe Shielding

Probe shielding is used to prevent or reduce the interaction of the probe's magnetic field with nonrelevent features in close proximity of the probe. Shielding could be used to reduce edge effects when testing near dimensional transitions such as a step or an edge. Shielding could also be used to reduce the effects of conductive or magnetic fasteners in the region of testing.

Eddy current probes are most often shielded using magnetic shielding or eddy current shielding. Magnetically shielded probes have their coil surrounded by a ring of ferrite or other material with high permeability and low conductivity. The ferrite creates an area of low magnetic reluctance and the probe's magnetic field is concentrated in this area rather than spreading beyond the shielding. This concentrates the magnetic field into a tighter area around the coil.

Eddy current shielding uses a ring of highly conductive but nonmagnetic material, usually copper, to surround the coil. The portion of the coil's magnetic field that cuts across the shielding will generate eddy currents in the shielding material rather than in the non-relevent features outside of the shielded area. The higher the frequency of the current used to drive the probe, the more effective the shielding will be due to the skin effect in the shielding material.

Probe Loading with Ferrite Cores

Sometimes coils are wound around a ferrite core. Since ferrite is ferromagnetic, the magnetic flux produced by the coil prefers to travel through the ferrite as opposed to the air. Therefore, the ferrite core concentrates the magnetic field near the center of the probe. This, in turn, concentrates the eddy currents near the center of the probe. Probes with ferrite cores tend to be more sensitive than air core probes and less affected by probe wobble and liftoff.

■ 〖Point 4〗 Coil Design

The most important feature in eddy current testing is the way in which the eddy currents are induced and detected in the material under test. This depends on the design of the probe. As discussed above, probes can contain one or more coils, a core and shielding. All have an important effect on the probe, but the coil requires the most design consideration.

A coil consists of a length of wire wound in a helical manner around the length of a former. The main purpose of the former is to provide a sufficient amount of rigidity in the coil to prevent distortion. Formers used for coils with diameters greater than a few millimeters (i.e. encircling and pancake coils), generally take the form of tubes or rings made from dielectric materials. Small-diameter coils are usually wound directly onto a solid former.

The region inside the former is called the core, which can consist of either a solid material or just air. When the core is air or a nonconductive material, the probe is often referred to as an air-core probe. Some coils are wound around a ferrite core which concentrates the the coil's magnetic field into a smaller area. These coils are referred to as "loaded" coils.

The wire used in an eddy current probe is typically made from copper or other nonferrous metal to avoid magnetic hysteresis effects. The winding usually has more than one layer so as to increase the value of inductance for a given length of coil. The higher the inductance (L) of a coil, at a given frequency, the greater the sensitivity of eddy current testing.

It is essential that the current through the coil is as low as possible. Too high a current may produce:

· A rise in temperature, hence an expansion of the coil, which increases the value of L.

· Magnetic hysteresis, which is small but detectable when a ferrite core is used.

〖Point 5〗 Impedance Matching

Eddy current testing requires us to determine the components of the impedance of the detecting coil or the potential difference across it. Most applications require the determination only of changes in impedance, which can be measured with a high degree of sensitivity using an AC bridge. The principles of operation of the most commonly used eddy current instruments are based on Maxwell's inductance bridge, in which the components of the impedance of the detecting coil, commonly called a probe, are compared with known variable impedances connected in series and forming the balancing arm of the bridge (Fig. 7–23). Refer back to Bridges.

The input to the bridge is an AC oscillator, often variable in both frequency and amplitude. The detector arm takes the form of either a meter or a storage cathode-ray oscilloscope, a phase-sensitive detector, a rectifier to provide a steady indication, and usually an attenuator to confine the output indication within a convenient range. Storage facilities are necessary in the oscilloscope in order to retain the signal from the detector for reference during scanning with the probe.

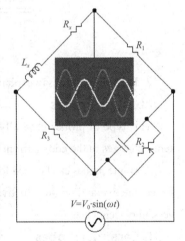

Fig. 7–23 Balancing Arm of the Bridge

The highest sensitivity of detection is achieved by properly matching the impedance of the probe to the impedance of the measuring instrument. Thus, with a bridge circuit that is initially balanced, a subsequent but usually small variation in the impedance of the probe upsets the balance, and a potential difference appears across the detector arm of the bridge.

Although the Maxwell inductance bridge forms the basis of most eddy current instruments, there are several reasons why it cannot be used in its simplest form (i.e. Hague, 1934), including the creation of stray capacitances, such as those formed by the leads and leakages to earth. These unwanted impedances can be eliminated by earthing devices and the addition of suitable impedances to produce one or more wide-band frequency (i.e. low Q) resonance circuits. Instruments having a wide frequency range (i.e. from 1 kHz to 2 MHz) may possess around five of these bands to cover the range. The value of the impedance of the probe is therefore an important consideration in achieving proper matching and, as a result, it may be necessary to change the probe when switching from one frequency band to another.

Put into Practice

1. Read the diagram to show which probes are used in Fig. 7–24.

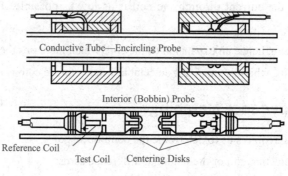

Fig. 7–24 ET Probes

2. The following essay explains what kinds of vortex probes there are.

Absolute Probes

This type of probes generally has a single test coil that is used to generate the eddy currents and sense changes in the eddy current field.

Absolute probes can be used for cracks detection, conductivity measurements, thickness measurements for conductive and non-conductive layers and liftoff measurements. They are widely used due to their versatility techniques.

Differential Probes

Differential probes have two active coils usually wound in opposition. When the two coils are over a flaw-free area of test sample, there is no differential signal appeared between the coils since they are both inspecting identical material. The differential signal appears, when one coil is over a crack and the other is over flaw-free area of testing material.

Differential probes have the advantage of being very sensitive to defects yet relatively insensitive to slowly varying properties such as electrical conductivity, magnetic permeability or temperature variations.

Reflection Probes

Probes of this type are often referred to as driver/pickup probes. They have at least two coils similar to a differential probe, but one coil is used to excite the magnetic field and eddy currents, and the other is used to sense changes in the test material. The advantages of reflection probes is that the driver and pickup coils can be separately optimized for their intended purpose. The driver coil can be made so as to produce a strong and uniform magnetic field around of the pickup coil, while the pickup coil can be desined very small so that it will be sensitive to very small defects.

7.4　Procedure Issues

■ 〖Point 1〗 Reference Standards

In eddy current testing, the use of reference standards in setting up the equipment is particularly important since signals are affected by many different variables and slight changes in equipment setup can drastically alter the appearance of a signal. As with most other NDT methods, the most useful information is obtained when comparing the results from an unknown object to results from a similar object with well characterized features and defects. In almost all cases, eddy current inspection procedures require the equipment to be configured using reference standards (Fig. 7-25).

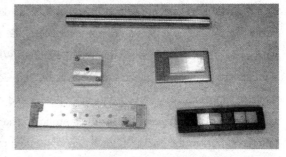

Fig. 7-25　Reference Standard

For crack detection, corrosion thinning and other material damage, reference standards are used to setup the equipment to produce a recognizable signal or set of signals from a defect or set of defects. In many cases, the appearance of a test signal can be related to the appearance of a signal from a known defect on the reference standard to estimate the size of a defect in the test component. Signals that vary signif icantly from the responses produced by the reference standard must be further investigated to the determine the source of the signal.

The reference standard should be of the same material as the test article. If this is not possible or practical, it should be of material that has the same electrical conductivity and magnetic permeability. Component features (material thickness, geometry, etc.) should be the same in the reference standard as those in the test region of interest. If the reference standard is the type with intentional defects, these defects should be as representative of actual defects in the test component as possible. The closer the reference standard is to the actual test component, the better. However, since cracks and corrosion damage are often difficult and costly to produce, artificial defects are commonly used. Narrow notches produced with electron discharge machining (EDM) and saw cuts are commonly used to represent cracks, and drilled holes are often used to simulate corrosion pitting.

Common eddy current reference standards include:

•Conductivity standards

- Flat plate discontinuity standards
- Flat plate metal thinning standards (step or tapered wedges)
- Tube discontinuity standards
- Tube metal thinning standards
- Hole (with and without fastener) discontinuity standards

〖Point 2〗 Signal Filtering

Signal filtering is often used in eddy current testing to eliminate unwanted frequencies from the receiver signal. While the correct filter settings can significantly improve the visibility of a defect signal, incorrect settings can distort the signal presentation and even eliminate the defect signal completely. Therefore, it is important to understand the concept of signal filtering.

Filtering is applied to the received signal and, therefore, is not directly related to the probe drive frequency. This is most easily understood when picturing a time versus signal amplitude display. With this display mode, it is easy to see that the signal shape is dependent on the time or duration that the probe coil is sensing something. For example, if a surface probe is placed on the surface of conductor and rocked back and forth, it will produce a wave like signal. When the probe is rocked fast, the signal will have a higher frequency than when the probe is rocked slowly back and forth. The signal does not need a wavelike appearance to have frequency content and most eddy current signals will be composed of a large number of frequencies. Consider a probe that senses a notch for 1/60th of a second. In a period of one second the probe could (in theory) go over the notch 60 times, resulting in the notch signal having a frequency of 60 Hz. But, imposed on this same signal, could be the signal resulting from probe wobble, electronic noise, a conductivity shift and other factors which occur at different frequencies.

The two standard filters found in most impedance plane display instruments are the "High Pass Filter" (HPF) and "Low Pass Filter" (LPF). Some instruments also have a "Band Pass Filter" (BPF), which is a combination high and low pass filter. Filters are adjusted in Hertz (Hz).

The HPF allows high frequencies to pass and filters out the low frequencies. The HPF is basically filtering out changes in the signal that occur over a significant period of time.

The LPF allows low frequency to pass and filters out the high frequency. In other words, all portions of the signal that change rapidly (have a high slope) are filtered, such as electronic noise.

In (Fig. 7-26), the gradual (low frequency) changes were first filtered out with a HPF and then high frequency electronic noise was filtered with a LPF to leave a clearly visible flaw indication. It should also be noted that since flaw indication signals are comprised of multiple frequencies, both filters have a tendency to reduce the indication signal strength.

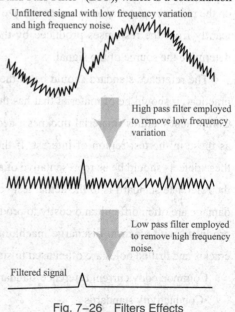

Fig. 7-26 Filters Effects

Additionally, scan speed must be controlled when using filters. Scan over a flaw too slow and the HPF might filter out the flaw indication. Scan over the flaw too fast and the LPF might eliminate the flaw indication.

If the spectrum of the signal frequency and the signal amplitude or attenuation are plotted, the filter responses can be illustrated in graphical form. Fig. 7–27 shows the response of a LPF of 20 Hz and a HPF of 40 Hz. The LPF allows only the frequencies in light gray to pass and the HPF only allow those frequencies in the dark gray area to pass. Therefore, it can be seen that with these settings there are no frequencies that pass (i.e. the frequencies passed by the LPF are filtered out by the HPF and visa versa).

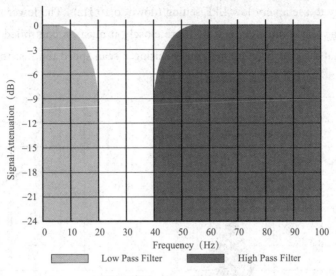

Fig. 7–27 Filter Settings

To create a window of acceptance for the signals, the filters need to overlap. In Fig. 7–28, the LPF has been adjusted to 60 Hz and the HPF to 10 Hz. The area shown in light gray is where the two frequencies overlap and the signal is passed. A signal of 30 Hz will get through at full amplitude, while a signal of 15 Hz will be attenuated by approximately 50%. All frequencies above or below the light gray area (the pass band) will be rejected by one of the two filters.

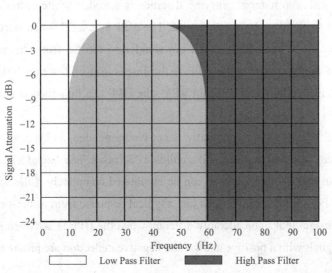

Fig. 7–28 Low and High Pass Filter

Use of Filters

The main function of the LPF is to remove high frequency interference noise. This noise can come from a variety of sources including the instrumentation and the probe itself. The noise appears as an unstable dot that produces jagged lines on the display as seen in the signal from a surface notch shown in the left image below. Lowering the LPF frequency will remove more of the higher frequencies from the signal and produce a cleaner signal as shown in Fig. 7-29. When using a LPF, it should be set to the highest frequency that produces a usable signal. To reduce noise in large surface or ring probes, it may be necessary to use a very low LPF setting (down to 10 Hz). The lower the LPF setting, the slower the scanning speed must be and the more closely it must be controlled. The image on the right shows a signal that has been clipped due to using a scan speed too fast for the selected HPF setting.

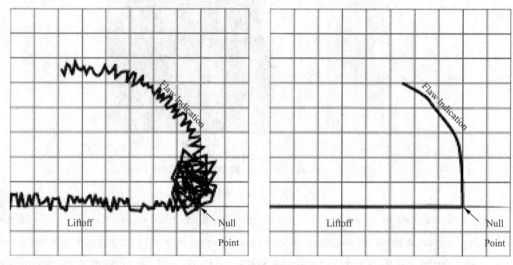

Fig. 7-29 HPF

The HPF is used to eliminate low frequencies which are produced by slow changes, such as conductivity shift within a material, varying distance to an edge while scanning parallel to it, or out-of-round holes in fastener hole inspection. The HPF is useful when performing automated or semiautomatic scans to keep the signal from wandering too far from the null (balance) point. The most common application for the HPF is the inspection of fastener holes using a rotating scanner. As the scanner rotates at a constant RPM, the HPF can be adjusted to achieve the desired effect.

Use of the HPF when scanning manually is not recommended, as keeping a constant scanning speed is difficult, and the signal deforms and amplitude decreases. The size of a signal decreases as the scan speed decreases and a flaw indication can be eliminated completely if the scan is not done with sufficient speed. In Fig. 7-30, it can be seen that a typical response from a surface notch in aluminum without HPF (left image) looks considerably different when the HPF is activated (right image). With the HPF, looping signals with a positive and similar negative deflection are produced on the impedance plane.

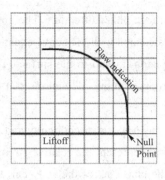 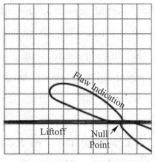 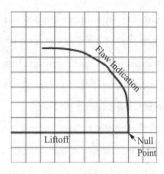

Fig. 7-30 Looping Signals

The use of a minimal HPF setting (1 or 2 Hz) may be used when manually scanning, provided the operator can largely control the scan speed and becomes familiar with the indication signal changes as scan speed is varied slightly. A good example of such an application would be the manual scan of the radius of a wheel that is rotated by hand, but the speed of rotation can be kept relatively constant.

〖Point 3〗 Phase Lag

Phase lag is a parameter of the eddy current signal that makes it possible to obtain information about the depth of a defect within a material. Phase lag is the shift in time between the eddy current response from a disruption on the surface and a disruption at some distance below the surface. The generation of eddy currents can be thought of as a time dependent process, meaning that the eddy currents below the surface take a little longer to form than those at the surface. Disruptions in the eddy currents away from the surface will produce more phase lag than disruptions near the surface. Both the signal voltage and current will have this phase shift or lag with depth, which is different from the phase angle discussed earlier. (With the phase angle, the current shifted with respect to the voltage.)

Phase lag is an important parameter in eddy current testing because it makes it possible to estimate the depth of a defect, and with proper reference specimens, determine the rough size of a defect. The signal produced by a flaw depends on both the amplitude and phase of the eddy currents being disrupted. A small surface defect and large internal defect can have a similar effect on the magnitude of impedance in a test coil. However, because of the increasing phase lag with depth, there will be a characteristic difference in the test coil impedance vector.

Phase lag can be calculated with the following equation. The phase lag angle calculated with this equation is useful for estimating the subsurface depth of a discontinuity that is concentrated at a specific depth. Discontinuities, such as a crack that spans many depths, must be divided into sections along its length and a weighted average determined for phase and amplitude at each position below the surface.

At one standard depth of penetration, the phase lag is one radian or 57°. This means that the eddy currents flowing at one standard depth of penetration (d) below the surface, lag the surface currents by 57°. At two standard depths of penetration ($2d$), they lag the surface currents by 114°. Therefore, by measuring the phase lag of a signal the depth of a defect can be estimated.

On the impedance plane, the liftoff signal serves as the reference phase direction. The angle between the liftoff and defect signals is about twice the phase lag calculated with the above equation. As mentioned above, discontinuities that have a significant dimension normal to the surface, will produce an angle that is based on the weighted average of the disruption to the eddy currents at the various depths along its length.

In the applet below, the relationship between the depth and dimensions of a discontinuity and the rotation produced on the impedance plane is explored. The red lines represent the relative strength of the magnetic field from the coil and the dashed lines indicate the phase lag of the eddy currents induced at a particular depth.

Put into Practice

1. Read the following essay to explain what components are required for the vortex detection heat exchange pipe.

Pipe inspection of Heat Exchangers

Pipe inspection of heat exchangers (Fig. 7-31) are suitable for checking various materials, like carbon steel, stainless steel, duplex, alloy or other conductive materials.

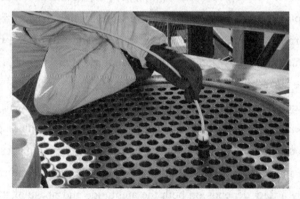

Fig. 7-31 Pipe Inspection of Heat Exchangers

Fig. 7-32 represents schematically the defect detection by eddy current. The two coils (red) were pulled over the defect and the typical signal appeared on the screen. The inner and outer defects have different signals depending on their deepness. This gives the operator the opportunity to decide what kind of defect is in the tube.

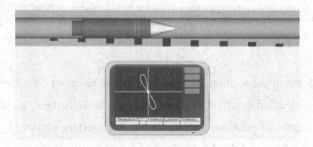

Fig. 7-32 The Typical Signal Appeared on the Screen

Eddy current testing is always a comparative measurement;therefore, EC-Works need:

• Inspection task based calibration pieces;

• Inspection task designed probes.

And very important:

• Eddy Current is not a leak-test system.

2. Read the following essay to explain what the vortex detection crack is through to show defects.

How does Eddy Current work ?

Fig. 7–33 represents schematically how does eddy current works. The coil causes a magnetic field, which initiate eddy current. If a defect appears in the effective area, the eddy current has to take a different way, this changes the impedance of the coil and the operator see the changes on his screen (Z_1–Z_2).

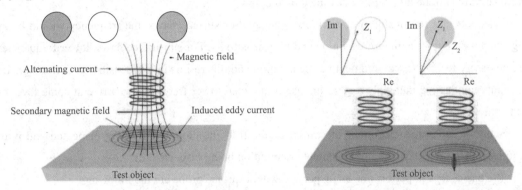

Fig. 7–33 Changes the Impedance of the Coil

7.5 Applications

〖Point 1〗 Surface Breaking Cracks(Fig. 7–34)

Eddy current equipment can be used for a variety of applications such as the detection of cracks (discontinuities), measurement of metal thickness, detection of metal thinning due to corrosion and erosion, determination of coating thickness, and the measurement of electrical conductivity and magnetic permeability. Eddy current inspection is an excellent method for detecting surface and near surface defects when

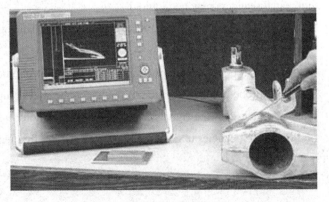

Fig. 7–34 Surface Breaking Cracks

the probable defect location and orientation is well known.

Defects such as cracks are detected when they disrupt the path of eddy currents and weaken their strength. The images show an eddy current surface probe on the surface of a conductive component. The strength of the eddy currents under the coil of the probe ins indicated by color. In the image, there is a flaw under the right side of the coil and it can be see that the eddy currents are weaker in this area.

Of course, factors such as the type of material, surface finish and condition of the material, the design of the probe, and many other factors can affect the sensitivity of the inspection. Successful detection of surface breaking and near surface cracks requires:

A knowledge of probable defect type, position and orientation.

Selection of the proper probe. The probe should fit the geometry of the part and the coil must produce eddy currents that will be disrupted by the flaw.

Selection of a reasonable probe drive frequency. For surface flaws, the frequency should be as high as possible for maximum resolution and high sensitivity. For subsurface flaws, lower frequencies are necessary to get the required depth of penetration and this results in less sensitivity. Ferromagnetic or highly conductive materials require the use of an even lower frequency to arrive at some level of penetration.

Setup or reference specimens of similar material to the component being inspected and with features that are representative of the defect or condition being inspected for.

The basic steps in performing an inspection with a surface probe are the following:

(1) Select and setup the instrument and probe.

(2) Select a frequency to produce the desired depth of penetration.

(3) Adjust the instrument to obtain an easily recognizable defect response using a calibration standard or setup specimen.

(4) Place the inspection probe (coil) on the component surface and null the instrument.

(5) Scan the probe over part of the surface in a pattern that will provide complete coverage of the area being inspected. Care must be taken to maintain the same probe-to-surface orientation as probe wobble can affect interpretation of the signal. In some cases, fixtures to help maintain orientation or automated scanners may be required.

(6) Monitor the signal for a local change in impedance that will occur as the probe moves over a discontinuity.

The applet below depicts a simple eddy current probe near the surface of a calibration specimen. Move the probe over the surface of the specimen and compare the signal responses from a surface breaking crack with the signals from the calibration notches. The inspection can be made at a couple of different frequencies to get a feel for the effect that frequency has on sensitivity in this application.

〖Point 2〗 Tube Inspection (Fig. 7-35 and Fig. 7-36)

Fig. 7-35   Tube Inspection   Fig. 7-36   Tube Inspection Instruments

Eddy current inspection is often used to detect corrosion, erosion, cracking and other changes in tubing. Heat exchangers and steam generators, which are used in power plants, have thousands of tubes that must be prevented from leaking. This is especially important in nuclear power plants where reused, contaminated water must be prevented from mixing with fresh water that will be returned to the environment. The contaminated water flows on one side of the tube (inside or outside) and the fresh water flows on the other side. The heat is transferred from the contaminated water to the fresh water and the fresh water is then returned back to is source, which is usually a lake or river. It is very important to keep the two water sources from mixing, so power plants are periodically shutdown so the tubes and other equipment can be inspected and repaired. The eddy current test method and the related remote field testing method provide high-speed inspection techniques for these applications.

A technique that is often used involves feeding a differential bobbin probe into the individual tube of the heat exchanger. With the differential probe, no signal will be seen on the eddy current instrument as long as no metal thinning is present. When metal thinning is present, a loop will be seen on the impedance plane as one coil of the differential probe passes over the flawed area and a second loop will be produced when the second coil passes over the damage. When the corrosion is on the outside surface of the tube, the depth of corrosion is indicated by a shift in the phase lag. The size of the indication provides an indication of the total extent of the corrosion damage.

A tube inspection using a bobbin probe is simulated below. Click the "null" button and then drag either the absolute or the differential probe through the tube. Note the different signal responses provided by the two probes. Also note that the absolute probe is much more sensitive to dings and the build up of magnetite on the outside of the tube than the differential probe is.

〖Point 3〗 Conductivity Measurements

One of the uses of eddy current instruments is for the measurement of electrical conductivity (Fig. 7-37). The value of the electrical conductivity of a metal depends on several factors, such as its chemical composition and the stress state of its crystalline structure. Therefore, electrical conductivity information can be used for sorting metals, checking for proper heat treatment, and inspecting for heat damage.

The technique usually involves nulling an absolute probe in air and placing the probe in contact with the sample surface. For non-magnetic materials, the change in impedance of the coil can be correlated directly to the conductivity of the material. The technique can be used to easily sort magnetic materials from nonmagnetic materials but it is difficult to separate the conductivity effects from the magnetic permeability effects, so conductivity measurements are limited to nonmagnetic materials. It is important to control factors that can affect the results such as the inspection temperature and the part geometry. Conductivity changes with temperature so measurements should be made at a constant temperature and adjustments made for temperature variations when necessary. The thickness of the specimen should generally be greater than three standard depths of penetration. This is so the eddy currents at the back surface of the sample are sufficiently weaker than the variations in the specimen thickness that are not seen in the measurements.

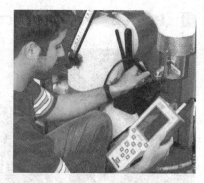

Fig. 7–37 The Measurement of Electrical Conductivity

Generally large pancake type, surface probes are used to get a value for a relatively large sample area. The instrument is usually setup such that a ferromagnetic material produces a response that is nearly vertical. Then, all conductive but non-magnetic materials will produce a trace that moves down and to the right as the probe is moved toward the surface. Think back to the discussion on the impedance plane and these type of responses make sense. Remember that inductive reactance changes are plotted along the y-axis and resistance changes are plotted in the x-axis. Since ferromagnetic materials will concentrate the magnetic field produced by a coil, the inductive reactance of the coil will increase. The effects on the signal from the magnetic permeability overshadow the effects from conductivity since they are so much stronger.

When the probe is brought near a conductive but nonmagnetic material, the coil's inductive reactance goes down since the magnetic field from the eddy currents opposes the magnetic field of the coil. The resistance in the coil increases since it takes some of the coil's energy to generate the eddy currents and this appears as additional resistance in the circuit. As the conductivity of the materials being tested increases, the resistance losses will be less and the inductive reactance changes will be greater. Therefore, the signals will be come more vertical as the conductivity increases, as shown in the image above.

To sort materials using an impedance plane device, the signal from the unknown sample must be compared to a signal from a variety of reference standards. However, there are devices available that can be calibrated to produce a value for electrical conductivity which can then be compared to published values of electrical conductivity in MS/m or percent IACS (International Annealed Copper Standard) . Please be aware that the conductivity of a particular material can vary significantly with slight variations in the chemical composition and, thus, a conductivity range is generally provided for a material. The conductivity range for one material may overlap with the range of a second material of interest, so conductivity alone can not always be used to sort materials. The electrical conductivity values for a variety of materials can be found in the material properties reference tables.

The following applet is based on codes for nonferrous materials written by Back Blitz from his book, "Electrical and Magnetic Methods of Non-destructive Testing", 2nd ed., Chapman & Hill (1997). The applet demonstrates how an impedance plane eddy current instrument can be used for the sorting of materials.

■ 〖Point 4〗 Thickness Measurements (Fig. 7-38) of Thin Material

Eddy current techniques can be used to perform a number of dimensional measurements. The ability to make rapid measurements without the need for couplant or, in some cases even surface contact, makes eddy current techniques very useful. The type of measurements that can be made include:

• Thickness of thin metal sheet and foil, and of metallic coatings on metallic and nonmetallic substrate

• Cross-sectional dimensions of cylindrical tubes and rods

• Thickness of nonmetallic coatings on metallic substrates

Fig. 7-38 Thickness Measurements

Corrosion Thinning of Aircraft Skins

One application where the eddy current technique is commonly used to measure material thickness is in the detection and characterization of corrosion damage on the skins of aircraft (Fig. 7-39). Eddy current techniques can be used to do spot checks or scanners can be used to inspect small areas. Eddy current inspection has an advantage over ultrasound in this application because no mechanical coupling is required to get the energy into the structure. Therefore, in multi-layered areas of the structure like lap splices, eddy current can often determine if corrosion thinning is present in buried layers.

Eddy current inspection has an advantage over radiography for this application because only single sided access is required to perform the inspection. To get a piece of film on the back side of the aircraft skin might require removing interior furnishings,

Fig. 7-39 The Detection and Characterization of Corrosion Damage the Skins of Aircraft

panels, and insulation which could be very costly. Advanced eddy current techniques are being developed that can determine thickness changes down to about three percent of the skin thickness.

Thickness Measurement of Thin Conductive Sheet, Strip and Foil

Eddy current techniques are used to measure the thickness of hot sheet, strip and foil in rolling mills, and to measure the amount of metal thinning that has occurred over time due to corrosion on fuselage skins of aircraft. On the impedance plane, thickness variations exhibit the same type of eddy current signal response as a subsurface defect, except that the signal represents a void of infinite size and depth. The phase rotation pattern is the same, but the signal amplitude is greater. In the applet, the liftoff curves for different areas of the taper wedge can be produced by nulling the probe in air and touching it to the surface at various locations of the tapered wedge. If a line is drawn between the end points of the liftoff curves, a comma shaped curve is produced. As illustrated in the second applet, this comma shaped curve is the path that is traced on the screen when the probe is scanned down the length of the tapered wedge so that the entire range of thickness values are measured.

When making this measurement, it is important to keep in mind that the depth of penetration of the eddy currents must cover the entire range of thicknesses being measured. Typically, a frequency is selected that produces about one standard depth of penetration at the maximum thickness. Unfortunately, at lower frequencies, which are often needed to get the necessary penetration, the probe impedance is more sensitive to changes in electrical conductivity. Thus, the effects of electrical conductivity cannot be phased out and it is important to verify that any variations of conductivity over the region of interest are at a sufficiently low level.

Measurement of Cross-sectional Dimensions of Cylindrical Tubes and Rods

Dimensions of cylindrical tubes and rods can be measured with either OD coils or internal axial coils, whichever is appropriate. The relationship between change in impedance and change in diameter is fairly constant, except at very low frequencies. However, the advantages of operating at a higher normalized frequency are two fold. First, the contribution of any conductivity change to the impedance of the coil becomes less important and it can easily be phased out. Second, there is an increase in measurement sensitivity resulting from the higher value of the inductive component of the impedance. Because of the large phase difference between the impedance vectors corresponding to changes in fill-factor and conductivity (and defect size), simultaneous testing for dimensions, conductivity and defects can be carried out.

Typical applications include measuring eccentricities of the diameters of tubes and rods and the thickness of tube walls. Long tubes are often tested by passing them at a constant speed through encircling coils (generally differential) and providing a close fit to achieve as high a fill-factor as possible.

An important application of tube-wall thickness measurement is the detection and assessment of corrosion, both external and internal. Internal probes must be used when the external surface is not accessible, such as when testing pipes that are buried or supported by brackets. Success has been achieved in measuring thickness variations in ferromagnetic metal pipes with the remote field technique.

Thickness Measurement of Thin Conductive Layers

It is also possible to measure the thickness of a thin layer of metal on a metallic substrate, provided the two metals have widely differing electrical conductivities (i.e. silver on lead where $s=67$

and 10 MS/m, respectively) . A frequency must be selected such that there is complete eddy current penetration of the layer, but not of the substrate itself. The method has also been used successfully for measuring thickness of very thin protective coatings of ferromagnetic metals (i.e. chromium and nickel) on non-ferromagnetic metal bases.

Depending on the required degree of penetration, measurements can be made using a single-coil probe or a transformer probe, preferably reflection type. Small-diameter probe coils are usually preferred since they can provide very high sensitivity and minimize effects related to property or thickness variations in the underlying base metal when used in combination with suitably high test frequencies. The goal is to confine the magnetizing field, and the resulting eddy current distribution, to just beyond the thin coating layer and to minimize the field within the base metals.

Put into Practice

1. Read the following essay to explain the advantages of vortex detection of thermal damage and material results.

Eddy current heat treat and material structure tests (Fig. 7–40) can be performed rapidly making the inspection easy to integrate into a production line. Other advantages include:

- Instantaneous responses
- Tests for multiple anomalies at once
- Green technology requiring no chemicals or couplants
- Reliable and reproducible results
- Easy integration into production lines for 100% part inspection rates

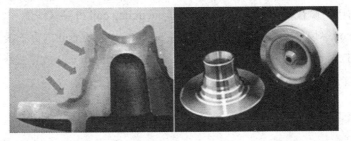

Fig. 7–40   Eddy Current Heat Treat and Material Structure Tests

2. Translate the following essay and answer the application scope of vortex detection.

Bar, Tube & Wire Inspection

Eddy current technology has been used to find flaws in bar, tube, and wire products for decades. The speed of response and repeatability of results make the inspection method perfect for high speed continuous flow production. Plus, materials do not have to be cleaned prior to inspection.

For finding medium to large flaws, a circular probe with multiple windings offers good results. If a flaw is discovered, our eddy current instrument sends commands to markers or cutters to highlight or remove defective product.

Fixturing is a critical factor to keep the material from wearing away the eddy current windings.

Our probes can be made from a few thousandths to many inches in diameter.

For finding very small flaws, such as pin holes in aluminum tubing, we use multiple coil flaw probes in specially designed fixturing. Our multi-channel eddy current instruments connect to multiple probes with overlapping coverage.

Other continuous flow testing applications include:
- Hydraulic tubing
- Medical tubing
- Air conditioning tubing

Words and Phrases

impedance analyzer ['ænəlaɪzə] n. 阻抗分析仪
multidimensional [ˌmʌltɪdaɪ'menʃənl] adj. 多维的，多面的
electromotive force (EMF) 电动势
solenoid ['səʊlənɔɪd] n. 螺线管
excitation coil 激励线圈
pickup coil 检测线圈
perpendicular to 垂直于
current-carrying wire 载流导线
electric charge 电荷
Biot-Savart law 毕奥–萨伐定律
unit direction vector 单位矢量
Faraday's law 法拉第定律

Lenz's law 楞次定律
self-inductance 自感
reactance [rɪ'æktəns] n. 电抗，阻抗
inductive reactance 感抗
mutual inductance 互感
secondary coil 次级线圈
primary (excitation) coil 初级线圈（激励线圈）
electrical conductivity 电导率
impedance plane diagram 阻抗平面图
angular separation 角距离
encircling coil 外穿式线圈
bobbin probe or coil 内穿式线圈
surface probe or coil 放置式线圈，点式线圈

A Sheet Work Manual

Pulsed Eddy Current Testing and Inspection

Determine the condition of your pipes and pressure vessels and monitor corrosion with our Pulsed Eddy Current services(Fig. 7–41).

Fig. 7–41 Pulsed Eddy Current Services

Assessing the condition of pipework and pressure vessels beneath insulation can be advantageous to plant operators. We apply Pulsed Eddy Current (PEC) technology to penetrate insulation and coating layers that are not magnetic and do not conduct electricity. We also use PEC technology to inspect and examine the low alloyed carbon steel substrate underneath metallic weather proofing sheeting and corrosion products. This approach allows our experts to easily and precisely test, inspect, measure and monitor the actual condition of your assets.

PEC technology does not require direct contact with a test object nor specific surface cleaning, making inspection fast and easy even at high temperatures and on offshore wells. Inspections can be conducted and corrosion can be monitored during operation to allow for planned maintenance and repairs to be scheduled and carried out at times optimal for your business.

We are licensed for operating PEC inspection equipment and possess the expert competences and technical experience necessary to conduct PEC testing. Make sure you are fully informed about the strength, safety and productivity of your assets at all times with a regular PEC testing and inspection strategy tailored to your requirements.

Look inside and learn the current condition and value of your assets!

Quick and reliable inspection at lower cost with state-of-the-art PEC technology.

The advantages of our pulsed eddy current technology are shown in Fig. 7–42.

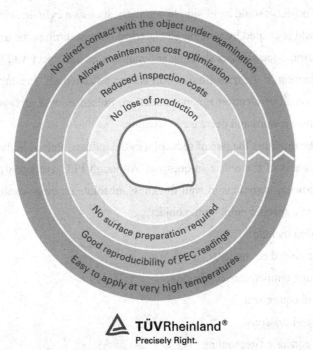

Fig. 7–42 The Advantages of Our Pulsed Eddy Current Technology

Benefit from the advantages of our Pulsed Eddy Current technology

Pulsed Eddy Current testing and inspection provides critical information on the actual condition of your assets, in order for you to monitor and manage any production or safety issues faster and easier than before. Because insulation materials need not be removed and surfaces require no particular preparation, overall inspection as well as underwater inspection costs are significantly reduced. In general, time allotted for PEC services is significantly less than conventional methods, making for fast and convenient in-operation testing and inspection with no loss in production.

Pulsed Eddy Current readings conducted many times at the same location can be reliably reproduced regardless of casing, coatings or insulation. PEC technology provides results with a plus/minus 10% accuracy for corrosion detection and a plus/minus 0.2% accuracy rate for corrosion monitoring. Moreover, Pulsed Eddy Current inspections can be successfully and easily carried out at temperatures ranging from −100 to 500 ℃ (−150 to 932 ℉).

Take advantage of our customized PEC services to ensure safety, productivity and profitability.

Pulsed Eddy Current technology for the testing and inspection of insulated pipes and vessels

Pulsed Eddy Current technology is based on electromagnetics and provides average wall thickness values over the probe footprint area. It measures and compares the percentage variation in average wall thickness throughout an object. Pulsed Eddy Current can be effectively applied for corrosion detection and monitoring on pipes and vessels made of carbon steel or low-alloy steel without making contact with the steel surface itself. PEC technology allows measurements to be made through insulation, concrete or corrosion barriers. We conduct Pulsed Eddy Current testing and inspection according to the following steps:

(1) An instrument probe is placed on the weather sheeting insulation, coating or rusted surface of a pipe or vessel. Test objects should be of simple geometry such as a cylinder, elbow or plate.

(2) A magnetic field is created by sending an electrical current through the transmitting coils of the probe. This field penetrates the weather sheeting and magnetizes the object wall.

(3) The electric current in the transmission coil is then switched off, causing a sudden drop in the magnetic field. As a result of electromagnetic induction, Pulsed Eddy Currents are generated in the object wall. They diffuse inward and decrease in strength.

(4) The PEC probe monitors the rate of reduction in the induced Pulsed Eddy Current to determine the average wall thickness of the object in question. Although PEC average wall thickness readings are relative values showing variations in wall thickness, absolute readings can be obtained with wall thickness calibration at a specific point on the object.

Areas of application include:
• Insulated and/or coated equipment
• Objects under high temperature conditions
• Heavily corroded equipment
• Offshore risers and caissons
• Objects behind concrete fireproofing
• Storage tank annular rings
• Bridges

Inspections can be conducted on objects with a wall thickness of 2 to 70 millimeters and a pipe diameter of more than 50 millimeters.

Cases of major safety incidents are shown in Table 7-1.

Table 7-1　Cases of Major Safety Incidents

Accident	Cases	Account	Link
	On October 1, 2017, an Air France Airbus A380 flying from Paris to Los Angeles disintegrated, its engine in theair during its transatlantic flight	If the blades on the engine break and fatigue cracks appear in the middle layer, the fan blades should be regularly ultrasonic phased array detection, and the non-destructive detection technology should be used to prevent them in advance	https://www.sohu.com/a/196127658_155305
	At 18:56 pm on April 6, 2015, video surveillance of Gulei emergency command center found that the PX project in Zhangzhou, Fujian Province exploded	Because the xylene device in the operation process of the pipeline welding port, due to false welding caused by the break, the leaking material was inhaled into the furnace, due to high temperature caused by explosion. Engineering Testing Co., Ltd. test results and accident investigation retest data do not match, suspected fraud	http://fj.sina.com.cn/news/m/2015-08-16/detail-ifxfxzzn7510225.shtml
	On October 25, 2010, the accident picture of the Arabian flying carpet equipment in Changlu farm showed that the welding part of the central shaft was broken and the big arm was out of control, resulting in the collapse of the equipment	Equipment a rotating arm of the center axis fracture, welding quality and welding process problems, under the role of long-term cross-strain stress, the shaft shoulder root fatigue, local cracks. Because this shaft is inside the equipment, not easy to disassemble, so the owner can not detect cracks in a timely manner, resulting in the crack surface continues to expand, until the central shaft can not withstand the load, a complete weld break	http://www.360doc.com/content/17/0411/12/15711419_644650343.shtml
	At about 5 am on April 12, 2011, the second suspender of the main span of Kuerle Kongque River Bridge on national highway 314 broke, causing the bridge deck to collapse about 10 meters long and 12 meters wide, resulting in traffic disruption	Corrosion of load-bearing steel cables is one of the main causes of collapse	https://news.qq.com/a/20110412/000697.htm

continued

Accident	Cases	Account	Link
	At 10:25 on November 22, 2013, the Donghuang oil pipeline of Sinopec pipeline storage and transportation branch in Qingdao Economic and Technological Development Zone of Shandong Province leaked crude oil into the municipal drainage culvert. Oil and gas accumulated in the culvert forming a closed space and exploded in case of sparks, causing 62 deaths and 136 injuries, with a direct economic loss of 751.72 million yuan	Pipeline corrosion at the intersection of oil pipelines and drainage channels is reduced, pipes are ruptured, oil spills are flowing into drainage channels and backsllowing to the road surface. The competent department of pipeline protection is not performing its duties well, and the investigation and control of safety hazards is not in-depth	http://xcepaper.zjol.com.cn/html/2014-02/28/content_6_3.htm

Words and Phrases

AC magnetic saturation 交流磁饱和
Absorbed dose 吸收剂量
Absorbed dose rate 吸收剂量率
Acceptance limits 验收范围
Acceptance level 验收水平
Acceptance standard 验收标准
Accumulation test 累积检测
Acoustic emission count (emission count) 声发射计数（发射计数）
Acoustic emission transducer 声发射换能器（声发射传感器）
Acoustic emission（AE）声发射
Acoustic holography 声全息术
Acoustic impedance 声阻抗
Acoustic impedance matching 声阻抗匹配
Acoustic impedance method 声阻法
Acoustic wave 声波
Acoustical lens 声透镜

Acoustic-ultrasonic 声-超声（AU）
Activation 活化
Activity 活度
Adequate shielding 安全屏蔽
Ampere turns 安匝数
Amplitude 幅度
Angle beam method 斜射法
Angle of incidence 入射角
Angle of reflection 反射角
Angle of spread 指向角
Angle of squint 偏向角
Angle probe 斜探头
Angstrom unit 埃（Å）
Area amplitude response curve 面积幅度曲线
Area of interest 评定区
Arliflcial discontinuity 人工不连续性
Artifact 假缺陷
Artificial defect 人工缺陷

Artificial discontinuity 标准人工缺陷
A-scan A 型扫描
A-scope; A-scan A 型显示
Attenuation coefficient 衰减系数
Attenuator 衰减器
Audible leak indicator 音响泄漏指示器
Automatic testing 自动检测
Autoradiography 自射线照片
Avaluation 评定
Barium concrete 钡混凝土
Barn 靶
Base fog 片基灰雾
Bath 槽液
Bayard-Alpert ionization gage B-A 型电离计
Beam 声束
Beam ratio 光束比
Beam angle 束张角
Beam axis 声束轴线
Beam index 声束入射点
Beam path location 声程定位
Beam path;path length 声程
Beam spread 声束扩散
Betatron 电子感应加速器
Bimetallic strip gage 双金属片计
Bipolar field 双极磁场
Black light filter 黑光滤波器
Black light; ultraviolet radiation 黑光
Blackbody 黑体
Blackbody equivalent temperature 黑体等效温度
Bleakney mass spectrometer 波利克尼质谱仪
Bleedout 渗出
Bottom echo 底面回波
Bottom surface 底面
Boundary echo (first) 边界一次回波
Bremsstrahlung 韧致辐射
Broad-beam condition 宽射束
Brush application 刷涂
B-scan presentation B 型扫描显示

B-scope; B-scan B 型显示
C-scan C 型扫描
Calibration instrument 设备校准
Capillary action 毛细管作用
Carrier fluid 载液
Carry over of penetrate 渗透剂移转
Cassette 暗合
Cathode 阴极
Central conductor 中心导体
Central conductor method 中心导体法
Characteristic curve 特性曲线
Characteristic curve of film 胶片特性曲线
Characteristic radiation 特征辐射
Chemical fog 化学灰雾
Cine-radiography 射线（活动）电影摄影术
Contact pads 接触垫
Circumferential coils 圆环线圈
Circumferential field 周向磁场
Circumferential magnetization method 周向磁化法
Clean 清理
Clean up 清除
Clearing time 定透时间
Coercive force 矫顽力
Coherence 相干性
Coherence length 相干长度（谐波列长度）
Coil test 测试线圈
Coil size 线圈大小
Coil spacing 线圈间距
Coil technique 线圈技术
Coil method 线圈法
Coil reference 线圈参考
Coincidence discrimination 符合鉴别
Cold-cathode ionization gage 冷阴极电离计
Collimator 准直器
Collimation 准直
Collimator 准直器
Combined colour contrast and fluorescent penetrant 着色荧光渗透剂

Compressed air drying 压缩空气干燥
Compressional wave 压缩波
Compton scatter 康普顿散射
Continuous emission 连续发射
Continuous linear array 连续线阵
Continuous method 连续法
Continuous spectrum 连续谱
Continuous wave 连续波
Contract stretch 对比度宽限
Contrast 对比度
Contrast agent 对比剂
Contrast aid 反差剂
Contrast sensitivity 对比灵敏度
Control echo 监视回波，参考回波
Couplant 耦合剂
Coupling 耦合
Coupling losses 耦合损失
Cracking 裂解
Creeping wave 爬波
Critical angle 临界角
Cross section 横截面
Cross talk 串音
Cross-drilled hole 横孔
Crystal 晶片
C-scope; C-scan C 型显示
Curie point 居里点
Curie temperature 居里温度
Curie (Ci) 居里
Current flow method 通电法
Current induction method 电流感应法
Current magnetization method 电流磁化法
Cut-off level 截止电平
Dead zone 盲区
Decay curve 衰变曲线
Decibel (dB) 分贝
Defect 缺陷
Defect resolution 缺陷分辨力
Defect detection sensitivity 缺陷检出灵敏度

Defect resolution 缺陷分辨力
Definition 清晰度
Definition, image definition 清晰度，图像清晰度
Demagnetization 退磁
Demagnetization factor 退磁因子
Demagnetizer 退磁装置
Densitometer 黑度计
Density 黑度（底片）
Density comparison strip 黑度比较片
Detecting medium 检验介质
Detergent remover 洗净液
Developer 显像剂
Developer station 显像工位
Developer agueons 水性显像剂
Developer dry 干显像剂
Developer liquid film 液膜显像剂
Developer nonaqueous（sus-pendible）非水（可悬浮）显像剂
Developing time 显像时间
Development 显影
Diffraction mottle 衍射斑
Diffuse indications 松散指示
Diffusion 扩散
Digital image acquisition system 数字图像识别系统
Dilatational wave 膨胀波
Dip and drain station 浸渍和流滴工位
Direct contact magnetization 直接接触磁化
Direct exposure imaging 直接曝光成像
Direct contact method 直接接触法
Directivity 指向性
Discontinuity 不连续性
Distance- gain- size-German AVG 距离 - 增益 - 尺寸（DGS 德文为 AVG）
Distance marker; time marker 距离刻度
Dose equivalent 剂量当量
Dose rate meter 剂量率计

Dosemeter 剂量计
Double crystal probe 双晶片探头
Double probe technique 双探头法
Double transceiver technique 双发双收法
Double traverse technique 二次波法
Dragout 带出
Drain time 滴落时间
Drain time 流滴时间
Drift 漂移
Dry method 干法
Dry powder 干粉
Dry technique 干粉技术
Dry developer 干显像剂
Dry developing cabinet 干显像柜
Dry method 干粉法
Drying oven 干燥箱
Drying station 干燥工位
Drying time 干燥时间
D-scope; D-scan D 型显示
Dual search unit 双探头
Dual-focus tube 双焦点管
Duplex-wire image quality indicator 双线像质指示器
Duration 持续时间
Dwell time 停留时间
Dye penetrant 着色渗透剂
Dynamic leak test 动态泄漏检测
Dynamic leakage measurement 动态泄漏测量
Dynamic range 动态范围
Dynamic radiography 动态射线透照术
Echo 回波
Echo frequency 回波频率
Echo height 回波高度
Echo indication 回波指示
Echo transmittance of sound pressure 往复透过率
Echo width 回波宽度
Eddy current 涡流
Eddy current flaw detector 涡流探伤仪

Eddy current testiog 涡流检测
Edge 端面
Edge effect 边缘效应
Edge echo 棱边回波
Edge effect 边缘效应
Effective depth penetration （EDP）有效穿透深度
Effective focus size 有效焦点尺寸
Effective magnetic permeability 有效磁导率
Effective permeability 有效磁导率
Effective reflection surface of flaw 缺陷有效反射面
Effective resistance 有效电阻
Elastic medium 弹性介质
Electric displacement 电位移
Electrical center 电中心
Electrode 电极
Electromagnet 电磁铁
Electro-magnetic acoustic transducer 电磁声换能器
Electromagnetic induction 电磁感应
Electromagnetic radiation 电磁辐射
Electromagnetic testing 电磁检测
Electro-mechanical coupling factor 机电耦合系数
Electron radiography 电子辐射照相术
Electron volt 电子伏特
Electronic noise 电子噪声
Electrostatic spraying 静电喷涂
Emulsification 乳化
Emulsification time 乳化时间
Emulsifier 乳化剂
Encircling coils 环绕式线圈
End effect 端部效应
Energizing cycle 激励周期
Equalizing filter 均衡滤波器
Equivalent 当量
Equivalent IQI. Sensitivity 像质指示器当量灵敏度

Equivalent nitrogen pressure 等效氮压
Equivalent penetrameter sensitivty 透度计当量灵敏度
Equivalent method 当量法
Erasabl optical medium 可探光学介质
Etching 浸蚀
Evaluation 评定
Evaluation threshold 评价阈值
Event count 事件计数
Event count rate 事件计数率
Examination area 检测范围
Examination region 检验区域
Exhaust pressure/discharge pressure 排气压力
Exhaust tubulation 排气管道
Expanded time-base sweep 时基线展宽
Exposure 曝光
Exposure table 曝光表格
Exposure chart 曝光曲线
Exposure fog 曝光灰雾
Exposure, radiographic exposure 曝光，射线照相曝光
Extended source 扩展源
Facility scattered neutrons 条件散射中子
False indication 假指示
Family 族
Far field 远场
Feed-through coil 穿过式线圈
Field, resultant magnetic 复合磁场
Fill factor 填充系数
Film speed 胶片速度
Film badge 胶片襟章剂量计
Film base 片基
Film contrast 胶片对比度
Film gamma 胶片 γ 值
Film processing 胶片冲洗加工
Film speed 胶片感光度
Film unsharpness 胶片不清晰度
Film viewing screen 观察屏

Filter 滤波器/滤光板
Final test 复探
Flat-bottomed hole 平底孔
Flat-bottomed hole equivalent 平底孔当量
Flaw 伤
Flaw characterization 伤特性
Flaw echo 缺陷回波
Flexural wave 弯曲波
Floating threshold 浮动阈值
Fluorescence 荧光
Fluorescent examination method 荧光检验法
Fluorescent magnetic particle inspection 荧光磁粉检验
Fluorescent dry deposit penetrant 干沉积荧光渗透剂
Fluorescent light 荧光
Fluorescent magnetic powder 荧光磁粉
Fluorescent penetrant 荧光渗透剂
Fluorescent screen 荧光屏
Fluoroscopy 荧光检查法
Flux leakage field 磁通泄漏场
Flux lines 磁通线
Focal spot 焦点
Focal distance 焦距
Focus length 焦点长度
Focus size 焦点尺寸
Focus width 焦点宽度
Focus (electron) 电子焦点
Focused beam 聚焦声束
Focusing probe 聚焦探头
Focus-to-film distance(f.f.d) 焦点-胶片距离（焦距）
Fog 底片灰雾
Fog density 灰雾密度
Footcandle 英尺烛光
Freguency 频率
Frequency constant 频率常数
Fringe 干涉带

Front distance 前沿距离
Front distance of flaw 缺陷前沿距离
Full-wave direct current（FWDC）全波直流
Fundamental frequency 基频
Furring 毛状痕迹
Gage pressure 表压
Gain 增益
Gamma radiography γ 射线透照术
Gamma ray source γ 射线源
Gamma ray source container γ 射线源容器
Gamma rays γ 射线
Gamma-ray radiographic equipment γ 射线透照装置
Gap scanning 间隙扫查
Gas 气体
Gate 闸门
Gating technique 选通技术
Gauss 高斯
Geiger-Muller counter 盖革 - 弥勒计数器
Geometric unsharpness 几何不清晰度
Gray (Gy) 戈瑞
Grazing incidence 掠入射
Grazing angle 掠射角
Group velocity 群速度
Half life 半衰期
Half-wave current (HW) 半波电流
Half-value layer (HVL) 半值层
Half-value method 半波高度法
Halogen 卤素
Halogen leak detector 卤素检漏仪
Hard X-rays 硬 X 射线
Hard-faced probe 硬膜探头
Harmonic analysis 谐波分析
Harmonic distortion 谐波畸变
Harmonics 谐频
Head wave 头波
Helium bombing 氦轰击法
Helium drift 氦漂移

Helium leak detector 氦检漏仪
Hermetically tight seal 气密密封
High vacuum 高真空
High energy X-rays 高能 X 射线
Holography (optical) 光全息照相
Holography acoustic 声全息
Hydrophilic emulsifier 亲水性乳化剂
Hydrophilic remover 亲水性洗净剂
Hydrostatic text 流体静力检测
Hysteresis 磁滞
IACS 国际船级社协会
ID coil ID 线圈
Image definition 图像清晰度
Image contrast 图像对比度
Image enhancement 图像增强
Image magnification 图像放大
Image quality 图像质量
Image quality indicator sensitivity 像质指示器灵敏度
Image quality indicator(IQI)/image quality indication 像质指示器
Imaging line scanner 图像线扫描器
Immersion probe 液浸探头
Immersion rinse 浸没清洗
Immersion testing 液浸法
Immersion time 浸没时间
Impedance 阻抗
Impedance plane diagram 阻抗平面图
Imperfection 不完整性
Impulse eddy current testing 脉冲涡流检测
Incremental permeability 增量磁导率
Indicated defect area 缺陷指示面积
Indicated defect length 缺陷指示长度
Indication 指示
Indirect exposure 间接曝光
Indirect magnetization 间接磁化
Indirect magnetization method 间接磁化法
Indirect scan 间接扫查

Induced field 感应磁场
Induced current method 感应电流法
Infrared imaging system 红外成像系统
Infrared sensing device 红外扫描器
Inherent fluorescence 固有荧光
Inherent filtration 固有滤波
Initial permeability 起始磁导率
Initial pulse 始脉冲
Initial pulse width 始波宽度
Inserted coil 插入式线圈
Inside coil 内部线圈
Inside-out testing 外泄检测
Inspection 检查
Inspection medium 检查介质
Inspection frequency/ test frequency 检测频率
Intensifying factor 增感系数
Intensifying screen 增感屏
Interal,arrival time （Δt_{ij})/arrival time interval （Δt_{ij}) 到达时间差（Δt_{ij})
Interfacc boundary 界面
Interface echo 界面回波
Interface trigger 界面触发
Interference 干涉
Interpretation 解释
Ion pump 离子泵
Ion source 离子源
Ionization chamber 电离室
Ionization potential 电离电位
Ionization vacuum gage 电离真空计
Ionography 电离射线透照术
Irradiance, E 辐射通量密度，E
Isolation 隔离检测
Isotope 同位素
K value K 值
Kaiser effect 凯塞效应
Kilo Volt (kV) 千伏特
Kiloelectron Volt (keV) 千电子伏特
Krypton 85 氪 85

L/D ratio L/D 比
Lamb wave 兰姆波
Latent image 潜像
Lateral scan 左右扫查
Lateral scan with oblique angle 斜平行扫查
Latitude (of an emulsion) 胶片宽容度
Lead screen 铅屏
Leak 泄漏孔
Leak artifact 泄漏器
Leak detector 检漏仪
Leak testtion 泄漏检测
Leakage field 泄漏磁场
Leakage rate 泄漏率
Leechs 磁吸盘
Lift-off effect 提离效应
Light intensity 光强度
Limiting resolution 极限分辨率
Line scanner 线扫描器
Line focus 线焦点
Line pair pattern 线对检测图
Line pairs per millimetre 每毫米线对数
Linear (electron) accelerator (LINAC) 电子直线加速器
Linear attenuation coefficient 线衰减系数
Linear scan 线扫查
Linearity (time or distance) 线性（时间或距离）
Linearity, anplitude 幅度线性
Lines of force 磁力线
Lipophilic emulsifier 亲油性乳化剂
Lipophilic remover 亲油性洗净剂
Liquid penetrant examination 液体渗透检验
Liquid film developer 液膜显像剂
Local magnetization 局部磁化
Local magnetization method 局部磁化法
Local scan 局部扫查
Localizing cone 定域喇叭筒
Location 定位
Location accuracy 定位精度

Location computed 定位，计算
Location marker 定位标记
Location upon delta T 时差定位
Location, clusfer 定位，群集
Location, continuous AE signal 定位，连续 AE 信号
Longitudinal field 纵向磁场
Longitudinal magnetization method 纵向磁化法
Longitudinal resolution 纵向分辨率
Longitudinal wave 纵波
Longitudinal wave probe 纵波探头
Longitudinal wave technique 纵波法
Loss of back reflection 背面反射损失，底面反射损失
Love wave 乐甫波
Low energy gamma radiation 低能 γ 辐射
Low-enerugy photon radiation 低能光子辐射
Luminance 亮度
Luminosity 流明
Lusec 流西克
Maga or million electron volts (MeV) 兆电子伏特
Magnetic history 磁化史
Magnetic hysteresis 磁性滞后
Magnetic particle field indication 磁粉磁场指示器
Magnetic particle inspection flaw indications 磁粉检验的伤显示
Magnetic circuit 磁路
Magnetic domain 磁畴
Magnetic field distribution 磁场分布
Magnetic field indicator 磁场指示器
Magnetic field meter 磁场计
Magnetic field strength 磁场强度 (H)
Magnetic field/field, magnetic 磁场
Magnetic flux 磁通
Magnetic flux density 磁通密度
Magnetic force 磁化力
Magnetic leakage field 漏磁场
Magnetic leakage flux 漏磁通

Magnetic moment 磁矩
Magnetic particle 磁粉
Magnetic particle indication 磁痕
Magnetic particle testing/magnetic particle examination 磁粉检测
Magnetic permeability 磁导率
Magnetic pole 磁极
Magnetic saturation 磁饱和
Magnetic storage meclium 磁储介质
Magnetic writing 磁写
Magnetizing 磁化
Magnetizing current 磁化电流
Magnetizing coil 磁化线圈
Magnetostrictive effect 磁致伸缩效应
Magnetostrictive transducer 磁致伸缩换能器
Main beam 主声束
Manual testing 手动检测
Markers 时标
MA-scope; MA-scan MA 型显示
Masking 遮蔽
Mass attcnuation coefficient 质量吸收系数
Mass number 质量数
Mass spectrometer （M.S.）质谱仪
Mass spectrometer leak detector 质谱检漏仪
Mass spectrum 质谱
Master/slave discrimination 主从鉴别
MDTD 最小可测温度差
Mean free path 平均自由程
Medium vacuum 中真空
Mega or million volt MV 兆伏特
Micro focus X-ray tube 微焦点 X 光管
Microfocus radiography 微焦点射线透照术
Micrometre 微米
Micron of mercury 微米汞柱
Microtron 电子回旋加速器
Milliampere (mA) 毫安
Millimetre of mercury 毫米汞柱
Minifocus X-ray tube 小焦点调射线管

305

Minimum detectable leakage rate 最小可探泄漏率
Minimum resolvable temperature difference (MRTD) 最小可分辨温度差
Mode 波形
Mode conversion 波形转换
Mode transformation 波形转换
Moderator 慢化器
Modulation transfer function (MTF) 调制转换功能
Modulation analysis 调制分析
Molecular flow 分子流
Molecular leak 分子泄漏
Monitor 监控器
Monochromatic 单色波
Movement unsharpness 移动不清晰度
Moving beam radiography 可动射束射线透照术
Multiaspect magnetization method 多向磁化法
Multidirectional magnetization 多向磁化
Multifrequency eddy current testiog 多频涡流检测
Multiple back reflections 多次背面反射
Multiple reflections 多次反射
Multiple back reflections 多次底面反射
Multiple echo method 多次反射法
Multiple probe technique 多探头法
Multiple triangular array 多三角形阵列
Narrow beam condition 窄射束
NC NC 网络计算机
Near field 近场
Near field length 近场长度
Near surface defect 近表面缺陷
Net density 净黑度
Net density 净(光学)密度
Neutron 中子
Neutron radiograhy 中子射线透照
Neutron radiography 中子射线透照术
Newton (N) 牛顿
Nier mass spectrometer 尼尔质谱仪

Noise 噪声
Noise equivalent temperature difference (NETD) 噪声当量温度差
Nominal angle 标称角度
Nominal frequency 标称频率
Non-aqueous liquid developer 非水性液体显像剂
Non-condensable gas 非冷凝气体
Non-destructive Examination（NDE）无损试验
Non-destructive Evaluation（NDE）无损评价
Non-destructive Inspection（NDI）无损检验
Non-destructive Testing（NDT）无损检测
Non-erasble optical data 可固定光学数据
Non-ferromagnetic material 非铁磁性材料
Non-relevant indication 非相关指示
Non-screen-type film 非增感型胶片
Normal incidence 垂直入射（亦见直射声束）
Normal permeability 标准磁导率
Normal beam method; straight beam method 垂直法
Normal probe 直探头
Normalized reactance 归一化电抗
Normalized resistance 归一化电阻
Nuclear activity 核活性
Nuclide 核素
Object plane resolution 物体平面分辨率
Object scattered neutrons 物体散射中子
Object beam 物体光束
Object beam angle 物体光束角
Object-film distance 被检体-胶片距离
Object-film distance 物体-胶片距离
Over development 显影过度
Over emulsfication 过乳化
Overall magnetization 整体磁化
Overload recovery time 过载恢复时间
Overwashing 过洗
Oxidation fog 氧化灰雾
Pair production 偶生成

Pair production 电子对产生
Pair production 电子偶的产生
Palladium barrier leak detector 钯屏检漏仪
Panoramic exposure 全景曝光
Parallel scan 平行扫查
Paramagnetic material 顺磁性材料
Parasitic echo 干扰回波
Partial pressure 分压
Particle content 磁悬液浓度
Particle velocity 质点（振动）速度
Pascal (Pa) 帕斯卡（帕）
Pascal cubic metres per second 帕立方米每秒 (Pa·m³/s)
Path length 光程长
Path length difference 光程长度差
Pattern 探伤图形
Peak current 峰值电流
Penetrameter 透度计
Penetrameter sensitivity 透度计灵敏度
Penetrant 渗透剂
Penetrant comparator 渗透对比试块
Penetrant flaw detection 渗透探伤
Penetrant removal 渗透剂去除
Penetrant station 渗透工位
Penetrant, water-washable 水洗型渗透剂
Penetration 穿透深度
Penetration time 渗透时间
Permanent magnet 永久磁铁
Permeability coefficient 透气系数
Permeability, a-c 交流磁导率
Permeability, d-c 直流磁导率
Phantom echo 幻象回波
Phase analysis 相位分析
Phase angle 相位角
Phase controlled circuit breaker 断电相位控制器
Phase detection 相位检测
Phase hologram 相位全息
Phase sensitive detector 相敏检波器

Phase shift 相位移
Phase velocity 相速度
Phase-sensitive system 相敏系统
Phillips ionization gage 菲利浦电离计
Phosphor 荧光物质
Photo fluorography 荧光照相术
Photoelectric absorption 光电吸收
Photographic emulsion 照相乳剂
Photographic fog 照相灰雾
Photostimulable luminescence 光敏发光
Piezoelectric effect 压电效应
Piezoelectric material 压电材料
Piezoelectric stiffness constant 压电劲度常数
Piezoelectric stress constant 压电应力常数
Piezoelectric transducer 压电换能器
Piezoelectric voltage constant 压电电压常数
Pirani gage 皮拉尼计
Pitch and catch technique 一发一收法
Pixel 像素
Pixel size 像素尺寸
Pixel, disply size 像素显示尺寸
Planar array 平面阵(列)
Plane wave 平面波
Plate wave 板波
Plate wave technique 板波法
Point source 点源
Post emulsification 后乳化
Post emulsifiable penetrant 后乳化渗透剂
Post-cleaning 后清除
Post-cleaning 后清洗
Powder 粉末
Powder blower 喷粉器
Powder blower 磁粉喷枪
Pre-cleaning 预清理
Pressure difference 压力差
Pressure dye test 压力着色检测
Pressure probe 压力探头
Pressure testing 压力检测

Pressure- evacuation test 压力抽空检测
Pressure mark 压痕
Pressure,design 设计压力
Pre-test 初探
Primary coil 一次线圈
Primary radiation 初级辐射
Probe gas 探头气体
Probe test 探头检测
Probe backing 探头背衬
Probe coil 点式线圈
Probe coil 探头式线圈
Probe coil clearance 探头线圈间隙
Probe index 探头入射点
Probe to weld distance 探头–焊缝距离
Probe/ search unit 探头
Process control radiograph 工艺过程控制的射线照相
Processing capacity 处理能力
Processing speed 处理速度
Prods 触头
Projective radiography 投影射线透照术
Proportioning probe 比例探头
Protective material 防护材料
Proton radiography 质子射线透照
Pulse 脉冲，脉冲波
Pulse echo method 脉冲回波法
Pulse repetition rate 脉冲重复率
Pulse amplitude 脉冲幅度
Pulse echo method 脉冲反射法
Pulse energy 脉冲能量
Pulse envelope 脉冲包络
Pulse length 脉冲长度
Pulse repetition frequency 脉冲重复频率
Pulse tuning 脉冲调谐
Pump-out tubulation 抽气管道
Pump-down time 抽气时间
Q factor Q 值
Quadruple traverse technique 四次波法

Quality (of a beam of radiation) 射线束的质
Quality factor 品质因数
Quenching 阻塞
Quenching of fluorescence 荧光的猝灭
Quick break 快速断间
Rad(rad) 拉德
Radiance, L 面辐射率，L
Radiant existence, M 辐射照度 M
Radiant flux; radiant power，ψ_e 辐射通量、辐射功率 ψ_e
Radiation 辐射
Radiation does 辐射剂量
Radio frequency (r-f) display 射频显示
Radio-frequency mass spectrometer 射频质谱仪
Radiograph 射线底片
Radiographic contrast 射线照片对比度
Radiographic equivalence factor 射线照相等效系数
Radiographic exposure 射线照相曝光量
Radiographic inspection 射线检测
Radiographic inspection 射线照相检验
Radiographic quality 射线照相质量
Radiographic sensitivity 射线照相灵敏度
Radiographic contrast 射线底片对比度
Radiographic equivalence factor 射线透照等效因子
Radiographic inspection 射线透照检查
Radiographic quality 射线透照质量
Radiographic sensitivity 射线透照灵敏度
Radiography 射线照相术
Radiological examination 射线检验
Radiology 射线学
Radiometer 辐射计
Radiometry 辐射测量术
Radioscopy 射线检查法
Range 量程
Rayleigh wave 瑞利波
Rayleigh scattering 瑞利散射

Real image 实时图像
Real-time radioscopy 实时射线检查法
Rearm delay time 重新准备延时时间
Rearm delay time 重新进入工作状态延迟时间
Reciprocity failure 倒易律失效
Reciprocity law 倒易律
Recording medium 记录介质
Recovery time 恢复时间
Rectified alternating current 脉动直流电
Reference block 参考试块
Reference beam 参考光束
Reference block 对比试块
Reference block method 对比试块法
Reference coil 参考线圈
Reference line method 基准线法
Reference standard 参考标准
Reflection 反射
Reflection coefficient 反射系数
Reflection density 反射密度
Reflector 反射体
Refraction 折射
Refractive index 折射率
Refrence beam angle 参考光束角
Reicnlbation 网纹
Reject; suppression 抑制
Rejection level 拒收水平
Relative permeability 相对磁导率
Relevant indication 相关指示
Reluctance 磁阻
Rem(rem) 雷姆
Remote controlled testing 机械化检测
Replenisers 补充剂
Representative quality indicator 代表性质量指示器
Residual magnetic field/field, residual magnetic 剩磁场
Residual technique 剩磁技术
Residual magnetic method 剩磁法

Residual magnetism 剩磁
Resistance (to flow) 气阻
Resolution 分辨力
Resonance method 共振法
Response factor 响应系数
Response time 响应时间
Resultant field 复合磁场
Resultant magnetic field 合成磁场
Resultant magnetization method 组合磁化法
Retentivity 顽磁性
Reversal 反转现象
Ring-down count 振铃计数
Ring-down count rate 振铃计数率
Rinse 清洗
Rise time 上升时间
Rise-time discrimination 上升时间鉴别
Rod-anode tube 棒阳极管
Roentgen (R) 伦琴
Roof angle 屋顶角
Rotational magnetic field 旋转磁场
Rotational magnetic field method 旋转磁场法
Rotational scan 转动扫查
Roughing 低真空
Roughing line 低真空管道
Roughing pump 低真空泵
Safelight 安全灯
Sampling probe 取样探头
Saturation 饱和
Saturation，magnetic 磁饱和
Saturation level 饱和电平
Scan on grid lines 格子线扫查
Scan pitch 扫查间距
Scanning 扫查
Scanning index 扫查标记
Scanning directly on the weld 焊缝上扫查
Scanning path 扫查轨迹
Scanning sensitivity 扫查灵敏度
Scanning speed 扫查速度

Scanning zone 扫查区域
Scattered energy 散射能量
Scatter unsharpness 散射不清晰度
Scattered neutrons 散射中子
Scattered radiation 散射辐射
Scattering 散射
Schlieren system 施利伦系统
Scintillation counter 闪烁计数器
Scintillator and scintillating crystals 闪烁器和闪烁晶体
Screen 屏
Screen unsharpness 荧光增感屏不清晰度
Screen-type film 荧光增感型胶片
SE probe SE 探头
Search-gas 探测气体
Second critical angle 第二临界角
Secondary radiation 二次射线
Secondary coil 二次线圈
Secondary radiation 次级辐射
Selectivity 选择性
Semi-conductor detector 半导体探测器
Sensitirity value 灵敏度值
Sensitivity 灵敏度
Sensitivity of leak test 泄漏检测灵敏度
Sensitivity control 灵敏度控制
Signal 信号
Signal gradient 信号梯度
Signal over load point 信号过载点
Signal overload level 信号过载电平
Signal to noise ratio 信噪比
Single crystal probe 单晶片探头
Single probe technique 单探头法
Single traverse technique 一次波法
Sizing technique 定量法
Skin depth 集肤深度
Skip distance 跨距
Skin effect 集肤效应
Skip point 跨距点

Sky shine (air scatter) 空中散射效应
Sniffing probe 嗅吸探头
Soft X-rays 软 X 射线
Soft-faced probe 软膜探头
Solarization 负感作用
Solenoid 螺线管
Soluble developer 可溶显像剂
Solvent remover 溶剂去除剂
Solvent cleaners 溶剂清除剂
Solvent developer 溶剂显像剂
Solvent remover 溶剂洗净剂
Solvent-removal penetrant 溶剂去除型渗透剂
Sorption 吸着
Sound diffraction 声绕射
Sound insulating layer 隔声层
Sound intensity 声强
Sound intensity level 声强级
Sound pressure 声压
Sound scattering 声散射
Sound transparent layer 透声层
Sound velocity 声速
Source 源
Source data label 放射源数据标签
Source location 源定位
Source size 源尺寸
Source-film distance 射线源–胶片距离
Spacial frequency 空间频率
Spark coil leak detector 电火花线圈检漏仪（电火花检测仪）
Specific activity 放射性比度
Specified sensitivity 规定灵敏度
Standard 标准
Standard 标准试样
Standard leak rate 标准泄漏率
Standard leak 标准泄漏孔
Standard tast block 标准试块
Standardization instrument 设备标准化
Standing wave; stationary wave 驻波

Step wedge 阶梯楔块
Step-wadge calibration film 阶梯楔块校准底片
Step-wadge comparison film 阶梯楔块比较底片
Step wedge 阶梯楔块
Stereo-radiography 立体射线透照术
Subject contrast 被检体对比度
Subsurface discontinuity 近表面不连续性
Suppression 抑制
Surface echo 表面回波
Surface field 表面磁场
Surface noise 表面噪声
Surface wave 表面波
Surface wave probe 表面波探头
Surface wave technique 表面波法
Surge magnetization 脉动磁化
Surplus sensitivity 灵敏度余量
Suspension 磁悬液
Sweep 扫描
Sweep range 扫描范围
Sweep speed 扫描速度
Swept gain 扫描增益
Swivel scan 环绕扫查
System exanlillatien threshold 系统检验阈值
System inclacel artifacts 系统感生物
System noise 系统噪声
Tackground, target 目标本底
Tandem scan 串列扫查
Target 靶
Television fluoroscopy 电视 X 射线荧光检查
Temperature envelope 温度范围
Tenth-value-layer (TVL) 十分之一值层
Test coil 检测线圈
Test quality level 检测质量水平
Test ring 试环
Test block 试块
Test frequency 试验频率
Test piece 试片
Test range 探测范围

Test surface 探测面
Testing, ulrasonic 超声检测
Thermal neutrons 热中子
Thermocouple gage 热电偶计
Thermogram 热谱图
Thermography, infrared 红外热成像
Thermoluminescent dosemeter (TLD) 热释光剂量计
Thickness sensitivity 厚度灵敏度
Third critical angle 第三临界角
Thixotropic penetrant 摇溶渗透剂
Thormal resolution 热分辨率
Threading bar 穿棒
Three-way sort 三挡分选
Threshold setting 门限设置
Threshold fog 阈值灰雾
Threshold level 挡值
Threshold tenet 门限电平
Throttling 节流
Through transmission technique 穿透技术
Through penetration technique 贯穿渗透法
Through transmission technique; transmission technique 穿透法
Through-coil technique 穿过式线圈技术
Throughput 通气量
Tight 密封
Total reflection 全反射
Total image unsharpness 总的图像不清晰度
Tracer probe leak location 示踪探头泄漏定位
Tracer gas 示踪气体
Transducer 换能器/传感器
Transition flow 过渡流
Translucent base media 半透明载体介质
Transmission 透射
Transmission densitometer 发射密度计
Transmission coefficient 透射系数
Transmission point 透射点
Transmission technique 透射技术

Transmittance, τ 透射率 τ
Transmitted film density 检测底片黑度
Transmitted pulse 发射脉冲
Transverse resolution 横向分辨率
Transverse wave 横波
Traveling echo 游动回波
Travering scan; depth scan 前后扫查
Triangular array 正三角形阵列
Trigger/alarm condition 触发/报警状态
Trigger/alarm level 触发/报警标准
Triple traverse technique 三次波法
True continuous technique 准确连续法技术
Trueattenuation 真实衰减
Tube current 管电流
Tube head 管头
Tube shield 管罩
Tube shutter 管子光闸
Tube window 管窗
Tube-shift radiography 管子移位射线透照术
Two-way sort 两挡分选
Ultra-high vacuum 超高真空
Ultrasonic leak detector 超声波检漏仪
Ultrasonic noise level 超声噪声电平
Ultrasonic cleaning 超声波清洗
Ultrasonic field 超声场
Ultrasonic flaw detection 超声探伤
Ultrasonic flaw detector 超声探伤仪
Ultrasonic microscope 超声显微镜
Ultrasonic spectroscopy 超声频谱
Ultrasonic testing system 超声检测系统
Ultrasonic thickness gauge 超声波测厚仪
Ultraviolet radiation 紫外辐射
Under development 显影不足
Unsharpness 不清晰
Useful density range 有效光学密度范围
UV-A A 类紫外辐射
UV-A filter A 类紫外辐射滤片
Vacuum 真空

Vacuum cassette 真空暗盒
Vacuum testing 真空检测
Van de Graaff generator 范德格拉夫起电机
Vapor pressure 蒸汽压
Vapour degreasing 蒸汽除油
Variable angle probe 可变角探头
Vee path V 形行程
Vehicle 载体
Vertical linearity 垂直线性
Vertical location 垂直定位
Visible light 可见光
Vitua limage 虚假图像
Voltage threshold 电压阈值
Voltage threshold 阈值电压
Wash station 水洗工位
Water break test 水膜破坏试验
Water column coupling method 水柱耦合法
Water column probe 水柱耦合探头
Water path; water distance 水程
Water tolerance 水容限
Water-washable penetrant 可水洗型渗透剂
Wave 波
Wave guide acoustic emission 声发射波导杆
Wave train 波列
Wave from 波形
Wave front 波前
Wave length 波长
Wave node 波节
Wedge 斜楔
Wet slurry technique 湿软磁膏技术
Wet technique 湿法技术
Wet method 湿粉法
Wetting action 润湿作用
Wetting agents 润湿剂
Wheel type probe; wheel search unit 轮式探头
White light 白光
White X-rays 连续 X 射线
Wobble 摆动

Wobble effect 抖动效应
Working sensitivity 探伤灵敏度
Wrap around 残响波干扰
Xeroradiography 静电射线透照术
X-radiation X 射线
X-ray controller X 射线控制器
X-ray detection apparatus X 射线探伤装置
X-ray film 射线胶片
X-ray paper X 射线感光纸
X-ray tube X 射线管
X-ray tube diaphragm X 射线管光阑
Yoke 磁轭
Yoke magnetization method 磁轭磁化法
Zigzag scan 锯齿扫查
Zone calibration location 时差区域校准定位
Zone location 区域定位

References

[1] Pipe Welding Handbook M-247 250B2012 – 07.
[2] ULTRASONIC TESTING TRAINING HANDBOOK.
[3]《美国无损检测手册》(磁粉卷).